I0505863

BY

JOSEPH MAZI

No Editors, No Reviewers.

All Thoughts and Words from Cover to Cover

Are Raw Expressions from the

Author Himself.

World's Greatest Riddle

Special Relativity

World's Greatest Riddle

Special Relativity

How to accept Relativity's discoveries WITHOUT abandoning our intuitive belief of an absolute reality.

Joseph Mazi

ISBN: 9798639854101
Publisher: Kindle Direct Publishing

Contact Author:

mazithoughts@yahoo.com

Author's YouTube Channel:

https://www.youtube.com/chan-nel/UCC4yTH7-h8ouTrbQsXnhy9Q

(Or search "Mazi Thoughts" in YouTube)

CONTENTS

Introduction

The idea of relative motion was adopted by the science com-munity in 1905, shortly after Albert Einstein developed the theory of Special Relativity, revolutionizing several aspects of physics. In it, he states that true motion does not exist—that everyone or -thing in the world may claim that they are motionless, while it is the universe around them that moves. The intriguing aspect of this notion is the insistence that as *everyone* may claim to be motionless at the same time, they would *all* be correct! How can everyone change locations relatively to one another without either of them truly moving? Special Relativity implies, for instance, that a lone object in space simply cannot claim to be in motion, given that there is no other thing to which it may compare its location. Moreover, it implies that one cannot *feel* motion; any force of deceleration or acceleration that we feel, may merely be a result of a physical or gravitational force exerted by our surroundings, and not necessarily an indication of movement. Thus, the theory claims that motion may only exist in a relative sense between

3

two or more comparable points of reference; hence the theory's name, "*Relativity.*"

This theory is quite counterintuitive—to such degrees that many feel as though it must be flawed, yet it is seemingly impossible to disprove... Much like a difficult riddle that presents an absurd notion; despite knowing that it just cannot be true, it is nevertheless difficult to explain why.

Many believe, since it has been reviewed and verified by countless physicists for the past one hundred and fifteen years, that the theory must hold validity; however, it has only been tested for mathematical flaws. Given that the mathematics are formulated *around* this concept of an alternate reality whereby motion is only relative, the mathematics are *only* coherent to that specifically assumed reality. If one were to prove that an absolute state of motion (the measurement of motion, independent of the perception or knowledge of any finite being or observer) *does* exist, the theory's entire formulaic structure would crumble.

As the theory of Special Relativity is inspected for error, its philosophically asserted premise that absolute motion does not exist (since it is not observable) flies under the radar if it is untrue; physicists have not discovered a means by which the true state of motion may be observed—if it even exists. The only hopes for doing so would lie in the discovery of a reference point in the universe which we may know for a fact to have a specific state of absolute (not relative) motion, with which we would finally be able to compare our motion and verify for certain that motion is indeed occurring in a universal manner.

As one may attempt to say that *light* has one constant speed through space to which we may compare our motion, Special Relativity throws in yet another counterintuitive claim

4

to fend off such an answer. The second and last rule that Einstein inserts into his twist of reality, is that the speed of light must be observed equally by all observers, regardless of what relative speed of motion anyone holds. Although this is dubbed the theory's *second* rule (or postulate), it is actually one in the same with the first rule. The first postulate claims that all may be considered motionless. In such context, given that we all observe the speed of light from a motionless standpoint, it is only logical to assume that we may all witness light's speed of motion equally. The reason that this postulate remains a theory, and not a fact, is that we have not yet discovered a sufficiently accurate means to verify with certainty that it is true.

Assuming that Einstein's two postulates are correct, it is then calculated that certain scenarios would display a difference in *time* between two objects as they travel at alternate speeds. As we watch, say, Andre travel at near the speed of light (relatively to our motionless standpoint), we will notice Andre's clock ticking much slower than ours, or perhaps barely at all. It is indeed true that objects observed at varying speeds seem to display varying times; clocks tick slower in satellites that orbit Earth at tremendous speeds. Rather than assuming that motion mechanically influences an object's particles to behave more slowly, however, Einstein claimed that time itself is altered.

This idea differs tremendously from the science community's former stance prior to the birth of Einstein's theory. Suppose that a giant clock stretches across the entire universe, ticking at one constant rate. Regardless by what speeds we each move as we walk on or fly around the clock, we would expect that time would remain the same for all of us—we would all be looking at the same clock when telling time. Einstein, though, countered this assumption with his belief that time, just as he said for motion, is relative to each observer's perspective.

Einstein presents a brilliant philosophical question of uncertainty (How do we know that we are actually moving?) to *assume* that absolute motion does not exist entirely. Hence, we have Special Relativity. The opposite and equally possible answer to this question is that motion is absolute, even for a lone object in space. In attempt to support and explain this answer, I will be referencing only the accepted laws of physics to break down Einstein's theory and to expose its fallacies. A counter-theory in support of the idea that motion is absolute shall be elucidated. Furthermore, time variations observed in objects at varying speeds and gravitational fields shall be explained from an objective and absolutive perspective of physics.

Einstein's declaration of a twisted reality is not the only of its kind. A trend in today's science of quantum physics is leaning towards the notion that an objective (or absolute) reality may not exist. An absolute reality is the idea that the existence of reality is not affected by, or dependent on, the perspective from which it is observed. For instance, the "double-slit experiment" suggests that photons (particles of light), as well as electrons, may behave in two separate manners simultaneously, both as waves and particles, and may simultaneously exist in multiple locations depending on the perspective from which the event is being observed. This concept is dubbed the "wave-particle duality." Physicists have arrived at such conclusions because they are unable to explain the experiment's results with the classical concepts of a distinct "particle" or "wave." Humanity's age-old intuitive concept that one thing cannot exist in two different places or do two different things at the same time is repeatedly challenged by quantum and theoretical physics.

As with Einstein's theory of Special Relativity, the indefinite conclusions following the double-slit experiment derive from a combined perspective of mathematics *and* philosophy. The gathered data and the mathematical procedures performed are by no means erroneous; physicists are obviously great at physics and organized with their experimental procedures—it's what they do! However, the philosophical and logical aspects of such theories are questionable.

It is for this reason that I will tackle the conceptual and factual aspects of Special Relativity and the double-slit experiment, in attempt to refine some of science's possible misconceptions. No conclusions are made without a strict structure of logic, and the consideration of existing physics laws. Scrutiny of the key concepts, definitions and properties involved in each topic illuminate the variables at hand, while logic governs the reasonably limited conclusions that may follow.

The purpose and goal of this work, aside from conveying different perspectives to the science community, is to demonstrate how one may scrutinize any aspect of reality in a strictly logical manner to bring about a definite truth. At times, some or all of reality is claimed to be relative or subjective, even by science. It is the author's goal to reveal that reality is absolute/objective in every way, whether you're exploring the depths of quantum physics, or even the seemingly ambiguous concepts of philosophy. For any matter at hand, there will always exist one correct answer, so long as the question is sufficiently specific; for something may not *be* if it is not a certain *way*.

World's Greatest Riddle

Special Relativity

Chapter 1:

When Fallacies Hide

At times, faulty logic may silently reside within an array of sophisticated concepts. The difficulty of distinguishing between fallacies and truthful ideas rises considerably as the topic increases in ambiguity. Illusions are accepted amidst a hazy setting, falsehoods are camouflaged by familiarity, and we progress with a train of thought, misguided by a wrongful path. We may set forth grueling efforts toward a work of art, only to later realize a flaw within the initial concept on which the entire study was founded. This dynamic yields to no level of expertise, for it is *deception* with which a hidden fallacy facilitates one's misjudgment. A mathematician may attempt to resolve an intricate problem of physics corresponding to real-life scenarios. In doing so, each procedure may be performed correctly; each number may be multiplied, divided, subtracted and added without error. However, if the relationship between the mathematical procedures used and the reality of the situation in question is ever so slightly misaligned, the validity of the conclusion in such circumstances may be sabotaged. It is no surprise that conceptual blunders may occur during the resolution of a complex problem. The mind's clarity of interpretation is prone to losing grip with reality as the problem becomes increasingly difficult to comprehend.

What may be surprising, though, is that even in the simplest of problems within which a faulty concept is well disguised, the cleverest of minds may be deceived; left unable to explain the absurd conclusions that follow. The following math riddle was developed in the early 1900s. Its solution requires a skill level of mathematics so rudimentary, that only the first-grade procedures of addition and subtraction of one- and two-digit numbers are necessary. Yet, somehow, even high school graduates and adults struggle to find an answer. The common name for this riddle is the "Missing Dollar Riddle." For convenience, it will be referred to as the hotel-riddle. *(Note that nowhere else in this book will the reader be asked to problem-solve, aside from this one and only mathematical example which will be used to make a strong point about the uncertain validity of physics theories in the science community.)*

HOTEL-RIDDLE:

Three friends decide to share a room at a hotel for one night. The clerk at the front desk tells them that the cost of the room is $30 per night. If one divides this cost evenly between the three friends, then each friend will need to pay $10 to the hotel. Immediately after they each make the payment of $10; the clerk realizes that he had misread the cost of the hotel room. The cost was actually $25 per night. To correct this mistake of the $5 overcharge, the clerk decides to reimburse the $5 to the three friends. Since the clerk knows that $5 cannot be evenly distributed amongst the three friends, he decides to give $3 of the $5 back to the three friends, by reimbursing each of them $1. The clerk is

now left with $2, which he has agreed to keep as a tip. Since each of the three friends have initially paid $10, and they were each given back $1, the three friends have now each spent a total of $9. This means that, altogether, the three friends have spent $27. These $27, together with the $2 that the clerk has kept as a tip, is a total of $29. Since the three friends have initially paid $30, and now there are only $29 to be accounted, how did the last dollar disappear?

The difficulty and confusion commonly experienced when tackling this problem is unapparent if the solution is clarified in advance. So, it is recommended that the reader attempts to solve it before reading the following answer.

The hotel-riddle suggests a counterintuitive notion that somehow, some money or value may disappear if it is rearranged or exchanged in a particular way. If one entrusts too much credit to a mathematical process and not enough attention to the sense of intuition, one is more likely to accept ridiculous conclusions as such offered by this riddle. It is intuitive knowledge that money may not simply disappear only because it is *moved around*. The reader knows both this and, since the problem is dubbed a "riddle," that there is a mistake within the riddle which must be corrected or explained. On the other hand, if a real-life math problem is much more complicated and its conclusion isn't as obviously incorrect, the error is often overlooked if intuition is ignored. In such circumstances, it is important to scrutinize the claim and strictly analyze all that is present, defining every miniscule meaning present in the big picture.

The hidden fallacy in this riddle suggests that the clerk's $2 tip is to be *added* to the $27, inferring that $29 represents all that is accounted for. Let us question the *meanings* of each mentioned entity for a moment. To solve this riddle, one must act as an accountant who is to track the money flow in this transaction. The accountant must analyze the three key identifiable occurrences… The money that was *initially spent* (1), the money that was *reimbursed* (2), and the money that was left over after reimbursement, in other words, the *total amount spent* (3). $30 were initially spent, and $3 were reimbursed, and so $27 are left over as the total amount spent. In what physical location are the $27 that were spent? $25 of that $27 are in the hotel's cash register, and $2 of that $27 are in the clerk's pocket as tip. Thus, the $2 are *already* included in the $27 amount. Yet, this riddle makes it seem as though the $2 should be *added* to the $27.

Riddles like this have been rightfully famous for a good reason. People of all ages, including adults who have graduated from high school, are baffled by a puzzle which only requires addition and subtraction of a couple of one- and two-digit numbers. How could an adult struggle to solve a first-grade math problem?

The Confusion

The difficulty that most face when encountering a problem of this nature derives not so much from the level of mathematics involved, but rather from the hidden fallacies camouflaged within the numbers and processes. One numerical digit may be interpreted in multiple, yet deceivingly similar ways. In the hotel-riddle, the "$2" that the clerk has kept as tip, may either be interpreted by the reader as "money spent," or "part of money spent." Just take a moment to look at the two of those expressions; the difference between them is ridiculously subtle! Both expressions technically bear the same meaning, while the second phrase is slightly more specific than the first. Nonetheless, this minor variance of interpretation will draw dramatically dissimilar results.

It would be correct to interpret the $2 as *"part"* of the money spent, and the $27 as the *"total"* amount spent (excluding the $3 reimbursement), which implies that the $2 is *inside* of the $27; deeming the summation of the two senseless. The riddle, though, *allows* the reader to interpret more vaguely, considering the $2 as simply "an amount that was spent." (Technically true! Though, not specific.) The $27 is also vaguely interpreted as "an amount that was spent." Given only these vague descriptions of the two numbers, it would not be illogical to sum the "spent $27" with the "spent $2" to conclude the "total amount spent." Generalization allows for the summation of the two numbers, while specification does not.

The hotel-riddle effortlessly deceives its audience by assigning the "$2" to a meaning that is correct, but vague enough to allow for its improper use to go unnoticed. Although the riddle doesn't specifically instruct the reader to interpret a vague meaning, the manner in which the $2 is used implies it. It's an easy fallacy to miss, given that both the $27 and the $2 belong to a similar category (money spent). This is exactly what allows a fallacy to successfully sell lies! When a vaguely expressed statement is true, its improper application to a more specific process or situation is often unnoticed. Thus, we see how easily even a slight play of words or concepts may derail a mathematician during the resolution of a rudimentary mathematical problem.

Although the individuals who were initially tricked by this problem didn't add, subtract, multiply or divide any digits incorrectly, they've accepted the incorrectly applied procedure of adding two numbers together that should not be summed.

Due to a slight misunderstanding of the concepts and meanings involved in the process, they didn't realize that adding the $2 to the $27 would bear no logical meaning to the reality of the situation.

Bullying an adult with a first-grade math problem is made possible with the aid of indistinctly misguiding concepts which hide within the problem. The hidden fallacy within a problem is the main pest. However, the complexity of the numbers and mathematics encompassing a problem helps to further disguise the pest. As the surrounding mathematics become more complex, more abstract, and more theoretical, bringing more mind-boggling concepts to the table for the reader to get lost into, the tiny hidden fallacies with which they are entangled become all the more undetectable.

Ascending Complexity

The gap in mathematical understanding between a first grader and an adult is very large, and yet adults struggle on this first-grade level math problem, only because of one simple fallacy. Determining what "two dollars" means in a simple exchange with a cashier somehow becomes a head-scratching test for high school graduates and adults who are skillful in algebra and flawless with addition and subtraction. This is what a fallacy can do. It will drag a college student back down to first grade. It can deem our brains incompetent for even a simple situation.

Accordingly, a question comes to mind. If a cleverly hidden fallacy in a first-grade math problem causes adults to struggle finding the solution, then who may be fooled by an arrangement of well-disguised fallacies in a problem involving the most sophisticated level of mathematics and physics that this world has ever seen? Whilst bearing in mind the derailment that fallacies are capable of causing (even in the simplest of concepts), we may expect such a riddle to fool even a genius mathematician in the distant future, who considers quantum physics a "rudimentary" level of mathematics.

Accepting Falsehood

From this perspective, we must understand that as numerous scientists approve of a highly intricate and counterintuitive theory, there is not necessarily a great deal of assurance that the theory is valid.

As the hotel-riddle claims that a dollar has disappeared, our knowledge that there is some kind of mistake comes from our intuitive understanding that money or objects cannot simply vanish. However, if the counterintuitive conclusion is not so clearly incorrect, and it is seemingly coherent with the surrounding information that *seems* correct (or cannot be disproven), then it is much more difficult to refute.

Complex theories involving abstract concepts and complicated supporting formulas have claimed counterintuitive notions such as "one's time slows down if one is moving faster." Discovering a fallacy in such complex and abstract context may seem nearly impossible. If by such circumstances a counterintuitive theory doesn't seem to contradict other aspects of reality, little resistance is applied. Although a theory's coherence with many aspects of reality doesn't indisputably prove a theory valid, it sure helps the science community accept it with open arms.

Let us assume for a moment that all of humankind lacks the rational understanding to solve the hotel-riddle. Let's say that the hotel-riddle is actually a theory which states that "some money may vanish from a certain perspective during a certain process of exchange." In other words, if an accountant tracks a particular sequence of transactions in a *certain manner*, some value of money will *actually* be lost, but only from that perspective. So, as the hotel riddle accounts the transactions by adding the $2 to the $27, one would say "yup, we lost a dollar if we're looking at it that way," instead of saying "we're looking at it incorrectly." As a hundred years pass by, no one is able to disprove this theory; namely, to find the theory's fallacy. At this point, given that a mathematical formula is presented by this theory (that is, the manner in which the values were exchanged and thereafter accounted), and that no one is able to refute its claim, society would be forced to accept the mathematically appointed notion that money simply is lost from a *certain* perspective, but not from all perspectives.

Thereafter, they may develop a mathematical formula which converts the riddle's accounting of exchanges to the accurate conclusion of $30, rather than $29. Supposing that such formulas are the *true* means of tracking the *mysterious* nature of such patterns of exchange, they would then use this coherent formula to apply it to all other similar situations. Different, or more complex patterns of exchange may require adaptations to the formula. Consequently, an entire field of mathematics is developed for simply accounting these types of transactions. If this field of study works, and it is coherent, then there would be no reason to abandon it, as it is both useful and coherent with reality.

Rather than realizing the conceptual issue with the hotel-riddle's accounting process and using that understanding to simply avoiding such mathematical errors in similar cases, they (the people of this imagined scenario) devise a mathematical formula to modify the incorrect method of accounting to *make* it work. This results in much more effort and computation than necessary, as well as an incomplete understanding of the true nature of reality.

It is in this manner that a counterintuitive and ultimately incorrect theory (or riddle) which is highly complicated, comparatively to our understanding, may be accepted as truth. This may even be the case with theories that remain accepted today.

Riddles of Science

Upon the mention of extremely complex and counterintuitive theories of physics, perhaps one that most commonly comes to mind is the famous theory of *Special Relativity*, theorized by Albert Einstein in 1905 (coincidentally around the same time that the hotel-riddle became popular).

Einstein, as a physicist and mathematician, is thought to have been one of science's greatest minds, as he single-handedly revolutionized the philosophy of physics. Having a tremendous expertise of his craft, Einstein developed a theory describing the nature of motion, time and space. Although his theory seems rather counterintuitive, it is of course supported by brilliant formulas which allow it to coherently align with the observable reality.

Einstein's theory of Special Relativity suggests that in physics, there is no difference between an object in constant motion (holding one unaltering speed) and a stationary (non-moving) object. A moving object may claim that it is stationary, while it is the universe around it that is moving. He goes on to explain that motion is only a relative function between two or more objects or points of reference; hence the name

'Relativity.' Relative motion is essentially a comparative distance between two or more objects as time goes by. Relativity claims that this distance may change despite absolute motion not existing. Only after consideration of this idealistic claim, is it then calculated that the rate of time or the length of space itself may be altered for an object as its velocity approaches the speed of light.

Einstein's theory has been accepted and approved by countless physicists around the world. Remind yourself by asking, though, that if high school graduates struggle to solve a first-grade math problem that misguides the solver with one simple fallacy, then what sort of physicist would be able to solve Einstein's complicated theoretical riddle, if it were unintentionally produced as such?

Considering that many physicists, if not most, have a difficulty with wrapping their minds around the concepts of this theory, finding a subtle fallacy as it hides within the meanings of its numbers and ideologies may seem virtually impossible. If Special Relativity does harbor a hidden fallacy, physicists today are a heck of a lot less equipped to solve Einstein's riddle than regular high school graduates are to solve a fallacious first-grade math problem. As the concepts and mathematics surrounding a fallacy become increasingly abstract and intricate, the clarity with which one may understand the true relationship between the concepts and reality diminishes, thus the difficulty of discovering the fallacy is dramatically increased.

One might say, then, that we have *time.* A high school graduate will *eventually* solve the first-grade math problem. A physicist's attempt to solve a riddle that is of *his own* level of mathematics, may be comparable to a first grader's attempt to solve the hotel-riddle. Though, yet again, a first grader may *eventually* solve it. Given 115 years since Einstein's theory was born, someone *should* have discovered its fallacy by now, if there is one.

The problem with refuting Einstein's riddle is that his claim is not observable. Special Relativity states that true motion does not exist, but that it only exists relatively. One way to look at this concept is by imagining a lone person in space who has nothing to compare his location to. He could be moving at a constant rate of 300 miles per hour without even noticing. There is no breeze to indicate that he may be moving, since there is no air. Even if there was, that could just mean that wind is blowing past him, but not necessarily that he is in motion.

In developing his theory, Einstein asked a brilliant philosophical question which elucidates the fact that science cannot *prove* whether or not we are moving. "What is motion? There is no definite evidence for it." This is true; however, he uses the fact that we cannot yet detect or observe true motion to claim that motion does not exist entirely. This isn't wrong by science. If science cannot prove or evidence an element, it will not assume its existence. If we are wrong, though, and true motion does exist, despite our lack of ability to detect it, then Einstein's theory may not be entirely true.

Despite having a century to review Einstein's theory for flaws, physicists have not yet discovered any fallacies within it (if there are any) because of science's *lack* of ability to observe Einstein's claim. No flaws were discovered within the mathematics of his theory, but they are only surely correct in light of the supposed reality that the theory claims we live in, whereby absolute motion doesn't exist. The fallacy in the hotel-riddle is at least observable. Special Relativity's faults may not only be cleverly hidden, but also completely invisible, because we cannot currently detect true motion, and so if Einstein's claim that true motion doesn't exist is a fallacy, we have no means by which to label it as such.

Considering that the world has yet to meet any physicists who have excelled far beyond Einstein's capabilities, it wouldn't be farfetched to say that Einstein's theory could be a fallacious riddle that remains unsolved until this day. We cannot call forth any advanced mathematicians from the distant future to help discover flaws in our science, though we do have an alternative means.

Conceptual Analysis

It may be assumed that the hotel-riddle cannot be solved without the skill set of adding or subtracting digits. Then again, if a detective, who can't add or subtract, was to question the statements and assumptions made within the riddle, could he debunk it? The detective may ask questions as follows... "What does this amount mean? Why is the riddle asking us to add this amount to that amount? Don't they fall under the same category? Is it okay to group them together? Wasn't the $2 defined as *money spent*? If you're calling the $27 *total money spent*, then why is the $2 *doing something* to the $27, isn't the $2 already in the same category as the $27?" Without attaining the skill of addition and subtraction, questioning the meanings of each element and grouping them together appropriately would help *anyone* to discover the fallacy.

Instead of diving into the mind-boggling formulas of a problem to search for flaws, we may bypass such complexities, looking within the concepts alone, before any formulas take place. Einstein's entire theory is based on one purely

22

conceptual (non-formulaic) assumption. If his assumption is refuted, the validity of the entire theory is compromised. The formulas supporting the theory are *dependent* on Einstein's assumption, as they only apply to the supposed reality whereby motion is *only* relative.

In the chapters following, after Einstein's theory and its key concepts are elaborated, potential fallacies within his theory are exposed. Later on, an explanation will be offered to reveal how Special Relativity may display accurate descriptions of reality, even though it is based on flawed reasoning.

Einstein's theory infers that motion and time are not absolute. Similarly, the double-slit experiment gave modern physicists leeway to claim that even the existence and location of particles are indefinite. This experiment and its resulting conclusions will also be scrutinized and tested for logical consistency.

As we uncover some of the fallacies in today's science, new theories sprout to fruition.

Chapter 2:

Special Relativity

The theory of Special Relativity explains the manner in which an observed object experiences *time* and *space* as its velocity is compared to the speed of light. Special Relativity is based on two concurring assumptions, or postulates. For convenience, in this book, the postulates shall be referred to as Postulate 1 and Postulate 2.

Postulate 1: Movement at a constant velocity feels no different from a motionless state (a smooth car ride feels the same as sitting in a parked car); therefore, there is no distinction in physics (or nature) between motion and stillness. A lone object in space may not claim its state of motion (whether it is moving or not), given that it has no point of reference (something to compare its location to). And so, no object has a true state of motion. If there *is* a point of reference for the object, the object cannot prove whether it is the reference point or the object itself that is moving. Thus, there is no object in the universe that anyone may know for sure is not moving (or that is moving at a certain speed). In other words, there are no points of reference that are known to have a certain state of motion that may be held relatively to all objects. Conclusively, since there is no distinction between constant motion and a motionless state, the true motion of a lone object does not exist; motion is only a relevant measure of distance between any two objects. That said, each object in the universe may claim that it is motionless while it is the rest of the universe that is moving around it. When motion changes (by accelerating, decelerating, or changing directions) the force that one feels as a result is not necessarily due to its own motion, but rather due to external influences. For example, "your spaceship isn't accelerating as you pull back into your seat, the universe is moving past your ship, and the universe's gravitational pull is tugging on you as it accelerates past you."

Postulate 2: The motion of light through space has a constant speed, approximately 300,000 km per second. Given that each object may claim to be motionless (since true motion does not exist) any object observing light will observe it from a still point of view. Thus, all objects will witness light to travel at the same speed. One may not *move* relatively to light in order to change one's perceived speed of light, because true movement does not exist.

In consideration of Postulates 1 and 2, Einstein deduced two features known as Time Dilation and Space Dilation. Next, Example 1 provides a common representation of Time Dilation. Although the characters and details of this example may vary between the many that are depicted around the world, it holds the same key principles.

EXAMPLE 1 - Special Relativity's Time Dilation:

Jack is in a space-ship, hovering above a giant platform in space as he watches it pass by under him.

A second observer, Eve, is standing on the platform. Eve watches Jack's ship pass by from her left to her right. She notices that the ship passes by at a constant speed, in a parallel direction to the platform.

Jack's ship emits a laser (a beam of light) straight down below him towards the platform. The laser leaves the ship, hits the platform, and reflects back up to the ship.

Jack and Eve each use their own stopwatch to measure the time that goes by from the moment the laser leaves the ship, to the moment it returns to the ship.

The following page describes the scenario from Jack's perspective.

<u>Jack's Perspective</u>: Per Postulate 1, from Jack's point of view, he may claim that he is motionless, while it is the platform below him (along with Eve) that is moving. Jack sees the laser travel straight down and back up to his stationary ship, while Eve moves from his left to his right. Figure 1 illustrates this scenario below.

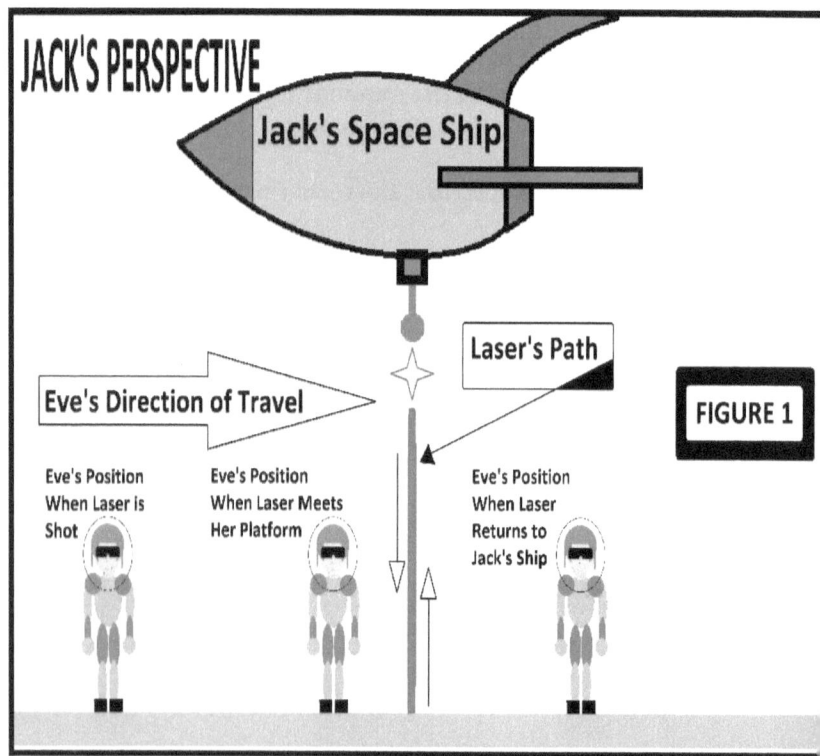

Eve's Perspective: Per Postulate 1, from Eve's point of view, she may claim that she and her platform are stationary, while it is the ship in front of her that is moving. As she watches the ship move horizontally from her left to her right, aside from noticing that the laser travels vertically, she must also notice that the laser is traveling horizontally along with the ship's horizontal movement, since the laser is aligned with the ship. Thus, Eve sees the laser travel a diagonal path. Figure 2 illustrates this scenario below.

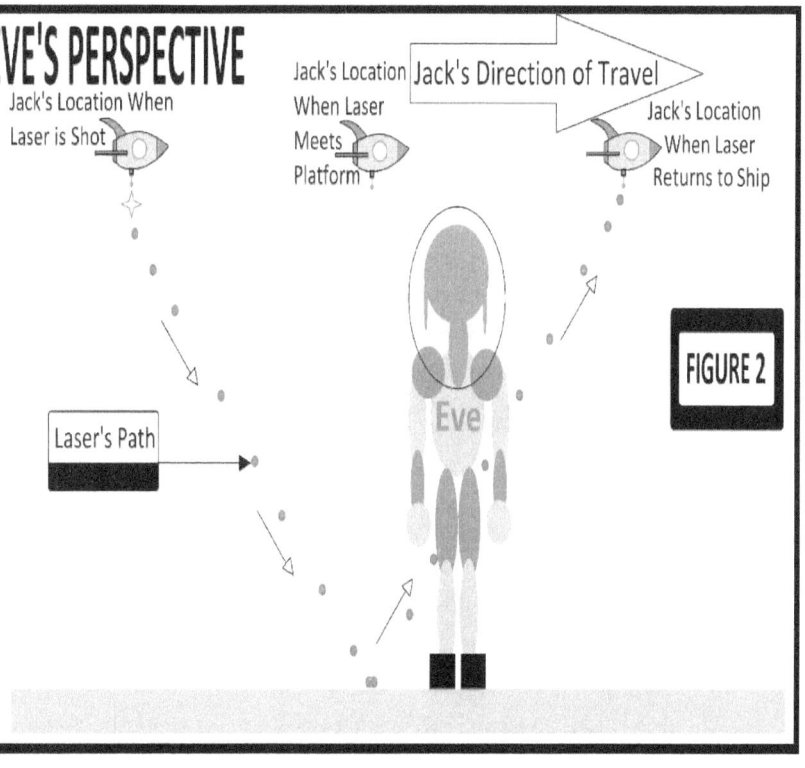

EVE'S PERSPECTIVE

Jack's Location When Laser is Shot

Jack's Location When Laser Meets Platform

Jack's Direction of Travel

Jack's Location When Laser Returns to Ship

FIGURE 2

Eve

Laser's Path

<u>Calculation</u>: Eve sees a greater total distance traveled by the laser, comparative to what Jack sees (the diagonal path is longer than the vertical path, given the same height). According to Postulate 2, the *speed* of light must be witnessed equally by all observers. Since Eve must see the same *speed* of light that Jack sees, and the laser travels a greater distance from Eve's perspective, it follows that it must take longer for the laser to return to the ship from Eve's perspective (laser travels same speed as it does in Jack's perspective, but it travels a longer distance, thus it takes longer to travel that distance). Since Eve measures a longer time comparatively to Jack for the same event, Eve's watch must be ticking faster than Jack's, as to accumulate more seconds before the laser reaches the ship.

<u>Conclusion</u>: Jack's watch must be running slower than Eve's. Thus, the theory of Special Relativity concludes that Jack's time must be slower than Eves time as he flies in his spaceship. Eve, in this case, will age faster than Jack, though each of them will feel as though their own times are running normally (from their own perspectives). The theory of Special Relativity uses this concept to imply that as an object's velocity is increased, its "time" runs slower.

Space dilation, the second feature Einstein deduced from his two Postulates, will be not be discussed, as it is merely a scenario similar to Example 1, using the same rules (postulates), which implies that one's relative length (aside from time) alters with varying speeds. The discussion of Time Dilation, alone, will be sufficient to uncover fallacies from the theory of Special Relativity.

Next, the laws of physics are utilized to refute Postulates 1 and 2 of Special Relativity, uncovering fallacies in Example 1.

Chapter 3:

Absolute Motion & Space

Upon comprehension of Time Dilation, many feel as though Special Relativity's assertions are absurd and nonsensical. It is indeed a counterintuitive theory. Many have attempted to discredit the theory; however, it continues to hold strong in the physics community. Upon reading the hotel-riddle, most know immediately that a dollar could not have disappeared, as such a supposition would defy logic. Nevertheless, most who attempt to solve it, struggle. There is a difference between knowing the answer (that the dollar is not missing) and *proving* the answer. Intuition may recognize potential fallacies or faults, though the true difficulty lies in discovering the factual reasoning to expose and explain them.

If Einstein's theory is invalid, then perhaps the most effective method of solving his riddle is met by shuddering the premise of his work. The validity of his entire theory is dependent on the accuracy of both of its postulates. If either of the theory's two postulates are compromised, the conceptual region of the theory, as it is understood today, may be par-tially or even completely erroneous.

Postulate 1's First Claim: Postulate 1 states that motion is only relative, claiming that an object, or even nature itself, may not know or feel the difference between stillness and mo-tion.

Postulate 1's Second Claim: Furthermore, it suggests that there exists no reference point in the universe that may cer-tainly indicate whether or not something is "truly moving."

Either one of these two statements may serve as a poten-tial vulnerability in Postulate 1's defense. Holding into consid-eration the latter claim that there is no reference point for ab-solute motion, a careful observation of the detailed events de-scribed by Example 1 draws four major fallacies, all of which ultimately evidence the universe's motionless reference point.

The first detail to stand out as strange resides in Example 1. When Eve assumes that Jack is moving horizontally (as she is motionless) why must she witness the laser to travel diago-nally? Perhaps, because of Jack's horizontal motion? If Jack is moving horizontally and throws a ball vertically, the ball will indeed travel both vertically *and* horizontally (in simpler terms, diagonally). Most people are familiar with this concept in physics, as the ball's horizontal momentum continues after it is released. The function of light, on the other hand, is of a different nature.

Light's Momentum

Suppose that a rifle in space is traveling in a horizontal di-rection (relatively to an observer) and shoots a bullet vertically. The bullet would simultaneously travel horizontally *and* vertically, which is essentially a diagonal movement. This occurs because, prior to being vertically shot, the *mass* of the bullet has a *velocity* in a horizontal direction. In other words, the bullet holds a horizontal *momentum* (momentum = mass x velocity) prior to being shot; and so, there is nothing to stop it from continuing that horizontal movement even well *after* departing from the rifle in a vertical direction. In the same scenario, if instead a photon (a particle of light) is emitted vertically from a horizontally moving flashlight, the photon would *only* travel vertically, without any horizontal movement, while the flashlight would continue to move horizontally. To understand why this occurs, the laws of physics regarding the characteristics of photons and *momentum* must be understood.

There are two reasons why a photon does not possess both vertical *and* horizontal motion when it is vertically emitted from a horizontally moving flashlight. The first reason is that, according to the laws of physics, a photon has absolutely *zero* mass. Without mass, a photon cannot absorb and accrue a physically driven momentum from the movement of its flashlight. Although a photon does have momentum, its momentum may only be calculated and influenced by its radioactive energy, wavelength, and its constant speed of light. The equation for the momentum of light (photon's energy divided by speed of light) simply does not involve a means by which light's momentum may acquire a movement in a direction as a result of the movement of its source. In other words, it does not move by the "physical" movement of the source; it only radiates as a wave. Thus, the horizontal movement of a flashlight should not affect the motion of a photon in the horizontal plane.

The second reason that a photon's direction of travel may not be influenced by its flashlight's motion is that a photon radiates away from the point of emission as a wave, it doesn't travel by force. A photon is not a physical object that is physically held or thrown by another object—it travels as do ripples through some sort of medium.

Conclusively, a flashlight in horizontal motion will not cause a vertically emitted photon to travel diagonally. Note that when a flashlight stays on, constantly generating light, one must not be fooled by the illusion whereby a *beam* of light "moves" horizontally as the flashlight from which it is emitted moves horizontally. It may appear so, only because the flashlight is constantly generating a renewed beam of light; as the flashlight's position is altered, a new beam of light is present in alignment with the flashlight's new location.

Relativity's 1st Fallacy: Light's Direction

The realization that a photon's *direction* of travel may not be influenced by the motion of its source exposes a fallacy in Example 1, which states that from Eve's perspective, since Jack's ship is considered to be in horizontal motion, the laser emitted vertically must have also traveled horizontally, creating a diagonal path. It is a direct violation of the laws of physics to assume that the vertically emitted laser (photon) may also travel horizontally as a result of the horizontal motion of the spaceship. Such an implication would suggest that the photons emitted must possess a horizontal momentum during emission, which is physically impossible due to the photon's massless content. This direct violation of the laws of physics immediately disqualifies the described calculations in Example 1 from taking place. If Eve assumes herself to be motionless, she should expect to see Jack's ship continue on to the right while its laser is left behind to make only a vertical path of travel, thus not returning to Jack's ship.

Eve may not and cannot, in any circumstance of Example 1, assume a diagonal motion of light, or a greater distance traveled by light. If Eve is moving horizontally while Jack remains motionless (meaning that the laser *does* return to Jack's ship), then she may experience an optical illusion as the laser *appears* to move diagonally; an optical illusion that Eve should not be fooled by given that she understands the laws of physics.

All considered, the laser in Example 1 should not have required more time to return to the ship from Eve's perspective, compared to Jack's perspective. Postulate 2 claims that the *speed* of light is constant for all observers, but it fails to acknowledge the uninfluenced *direction* of light by the motion of its source. If Jack instead shoots a *bullet* down vertically from his spaceship, then it would be impossible for either of the observers to determine which of them are certainly in motion; the bullet would remain horizontally aligned with Jack's motion whether he is or isn't in motion. Light, however, travels independently from the motion of its source; thus, Jack's true horizontal state of motion ought to be immediately exposed as it is compared to the horizontally motionless position of the laser's path of travel.

One opposing defense to this argument with which Special Relativity may respond, is the claim that "while Eve assumes that Jack is in horizontal motion, Jack's laser must still appear to be horizontally aligned with him only because Jack assumes himself to be motionless, meaning that Jack must see his laser to be aligned with his ship, thus so too should Eve."

The discrepancy with such logic is that it improperly *mixes* circumstances that do not belong together. In the scenario whereby Eve is considered to be motionless, Eve may ONLY assume *herself* to be motionless. She may not assume *both* of them to be motionless simultaneously, as that would result in no relative motion whatsoever. It would be illogical and incorrect to state that as Eve assumes herself to be motionless, she may *also* assume that Jack is motionless *only* when asking whether or not the laser should be aligned to Jack, but not while considering other aspects such as their relative position.

Given that light's direction may not be influenced by the motion of its source, Eve should expect, if she knows the correct laws of physics, that if Jack is truly in motion while Eve is truly motionless, Jack will continue to move on to Eve's Right side while his laser remains in front of Eve, as it follows a vertical (absolutely not diagonal) path from which point it was emitted.

One must also make note of another important fact in the case whereby Jack's laser may be traveling a diagonal path. For a diagonal path of light to have occurred, upon emission, the laser must have been *pointed* at an angle away from his ship, rather than downwards in a vertical direction. Light cannot travel sideways; having only one constant speed, *all* 300,000 km p/s of that speed is distributed *only* in one direction. Thus, to suggest that the laser travels diagonally from the point from which it was emitted, would imply that the ship is *pointing* the laser in that direction, altering the dynamics of the experiment.

Relativity's 2nd Fallacy: Geometry

EVEN IF one were to assume that a vertically emitted photon may move horizontally by influence of its source's horizontal motion, Example 1 yet draws another illogical assumption. An unforgivable fallacy lies in the claim that although the path of the photon was elongated, as it was horizontally influenced by Jack's horizontal motion, the photon CONTINUES to travel at the same constant speed! This is geometrically incorrect.

Suppose that a motionless rifle consistently shoots bullets through space at 16 miles per minute. If the rifle were to move horizontally at 10 miles per minute while shooting the bullet vertically, the bullet would then travel diagonally at a speed of roughly 18.87 miles per minute [square root of (vertical squared + horizontal squared)]. Geometrically, if a motion of *anything* (with or without mass) is a constant velocity in a vertical direction, and distance of travel is *added* horizontally without applying any resistance to the vertical motion, the total velocity of travel may *only* increase. Any distance traveled (per time) IS speed. To suggest that distance traveled-per-time may be added (in this case horizontally) without increasing speed, is to say that SPEED may be added without adding speed. The only reason that the path of light is elongated from Eve's perspective is that a horizontal distance of travel is added. If motion per distance is applied, then speed is applied, therefore increased. Thus, light's speed must only increase if its motion may truly be influenced on a horizontal plane.

If one truly believes that Jack's horizontal motion may influence the horizontal direction of the photon's travel, then one must accept that Postulate 2 (light speed is constant) is incorrect and that the speed of light has increased. This would dismiss the conclusion of Time Dilation. In Example 1, if light's speed was appropriately increased as a result of the added horizontal motion, then Eve would not have timed the event differently from Jack.

An alternative option is to consider that Postulate 2 *is* correct, in which case the vertical speed of light must somehow slow down as the horizontal speed is increased, in order for light to retain its total constant speed of 300,000 km per second. In such a case, the strange alteration of light's vertical speed may then be used as an indicator for Jack to determine whether or not he is in motion. If traveling horizontally causes Jack's laser to slow down vertically, then he may conclude that he is in motion. Whenever Jack observes that his laser does not return to him at a normal speed of light, (given that we assume Postulate 2 to be correct) he must assume that his laser has a horizontal direction that he is not optically capable of detecting; suggesting a mere optical illusion, but by no means witnessing a different speed of light. Thus, the *slowing down* of a photon's vertical travel as a result of the source's horizontal motion should be just as apparent for Jack as it was for Eve, and this would serve as an indicator for absolute motion! That's right, a reference for absolute motion is found within Special Relativity's very own example, without changing any of its rules! This right here is major fallacy, given that a contradiction is found within Special Relativity's own program.

In either case, whether an added horizontal motion of light increases its total velocity or slows down its vertical velocity, Time Dilation does not occur, mathematically speaking.

Relativity's 3rd Fallacy: Light's Second Dimension

Independently of the first two fallacies that were exposed in the theory of Special Relativity, there is yet a third one to confront. What do physicists know about the motion of light? They know that its speed is roughly 300,000 km per second.

This is true; however, why do physicists only acknowledge one speed of light? Like anything, light's velocity must be identified for two dimensions; the direction of travel, and the direction perpendicular to its direction of travel. If light is traveling *only* vertically, we know that its horizontal velocity must be absolutely 0 km per second. Why then, does Special Relativity state that there is no reference point to determine actual motion, when it knows that the motion of light must *always* be zero perpendicularly to its direction of travel?

The only thing in the universe that is thought to effect light's motion perpendicularly, is an outrageously enormous effect of gravity. By *enormous*, we're talking black holes and giant stars; and even they must be *very* close to light to have any effect on its direction of travel. So, if an observer does not see any giant stars or black holes nearby, then he or she may assume that photons traveling nearby are perpendicularly motionless. Thus, any motion that one notices relatively to the perpendicularly motionless path of a photon, *is* a depiction of an absolute state of motion.

Relativity's 4th Fallacy: Postulate 2 vs. Doppler Effect

Postulate 2 of Special Relativity states that the speed of light is observed equally by all observers, regardless of each individual's state of motion. This would mean that, as Bob is traveling forward at 50 km/s, and Bob emits a light in a forward direction, to Bob, the light would appear to be traveling away from him at the speed of light; not 50 km/s *slower* than the speed of light. Bob's speed relatively to light may not change, according to Postulate 2. Even if Bob travels forward at a million miles per hour, light would nonetheless move away from him at light-speed. This concept contradicts the

Relativity's 4th Fallacy: Postulate 2 vs. Doppler Effect

doppler effect that is proven to occur in light. The doppler effect is a change in the wave-frequency of light as the source of light moves towards or away from the direction of light's travel.

As you point a flashlight forward and turn it on, the flashlight is constantly generating light waves forward at a certain frequency. The gap in between each light wave emitted is the wave length. If the waves are generated at a higher frequency, meaning more times per second, the distance in between the waves will be shorter. If the waves are generated at a lower frequency, the distance in between the waves will be longer, simply because more time has elapsed in between the generation of each wave. Given that the flashlight emits waves at a consistent rate, it has been verified in experiments that if you were to move forward at any velocity while shining a light forward, the light will have a higher frequency, and it would have a lower frequency if you were to travel backwards. Why does this occur?

Let's play the scenario in super slow motion, only we will not follow the laws of Special Relativity, just to see what happens. You're pointing the flashlight forward. The first wave is emitted as it travels away from you at light-speed, and it covers a certain distance before the second wave is emitted. Now there is a certain distance in between the two traveling waves; let's call that distance one meter, just to keep things simple.

You start this experiment over, but this time you move forward at half the speed of light as you shine the flashlight forward. The first wave is emitted, traveling forward at the speed of light. This time, you're chasing after the first wave at half the speed of light, so the wave will only gain half as much distance away from you by the time the second wave is emitted, compared to when you weren't moving. And so, when you move forward as you point a flashlight forward, the wave lengths will be shorter and closer together, thus having a higher frequency.

In this experiment, if you were to instead travel backwards while pointing the flashlight forward, you would actually let the first wave gain *even more* distance away from you before the second wave is emitted, causing the wave lengths to be much longer and further apart, resulting in a lower frequency.

This objectively explains why the doppler effect occurs with light, and proves that light does not move in respect to its source as if the source is motionless. The doppler effect of light may only occur if the source of light is allowed to chase after or run away from said light. Bearing this concept in mind, if you were to hypothetically travel forward at the speed of light, and you shine a flashlight forward, the light would not gain any distance away from you, rather it would seem motionless relatively to yourself, since you and the light are both traveling forward at the same speed of light.

Special Relativity, on the other hand, claims that light would actually move away from you at the speed of light in this scenario, not to say that it is traveling twice the speed of light, but instead to say that you were never really traveling in the first place, because Special Relativity claims that from your perspective, you are never truly in motion. Very weird, huh? But if Special Relativity is correct, then the doppler effect could not be possible with light—and yet... it is.

Counter-Theory to Special Relativity

At least until the publication of this book, physicists have yet to discover a stationary reference point which indicates the true state of motion of all objects. Light was never thought to have been such a reference point, because its rate of travel is said to be witnessed equally by all eyes observing it, per Postulate 2 of Special Relativity. Physicists have failed to realize, however, that there are two dimensions of a photon's motion; the direction in which it travels, and the direction perpendicular to its travel. Thus, a universally motionless reference point is now indicated, theorizing a new concept in physics.

MAZI'S 1st THEORY: PHOTONIC REFERENCE POINT

Given that a photon is massless, it may not accrue momentum from the physical motion of its source. Therefore, a photon's velocity in the perpendicular direction relatively to its direction of emission must be absolutely zero. This motionless dimension of a photon serves as a reference point which indicates the stationary framework of space and at the very least, the absolute motion of any object or thing. Leaving the direction of emission aside, the direction of travel, too, indicates a photon's motionless state perpendicularly to said direction.

To better understand how a photon's path of travel may be used as a stationary reference point, consider the following simple scenario. You are standing on the floor. There is a vertical pole in front of you which represents the path by which the photon is traveling. The surface of the ground and the pole are perpendicular to one another. This means that if you jump upwards, you are moving in a parallel direction, relatively to the pole. Any direction that you may move as you walk along the floor, however, is considered perpendicular to the vertical

42

pole. As you walk to your left, right, front or rear, you will be able to identify your change of location by comparing your position to that of the stationary pole. With this method, you may only describe your direction of travel along the floor, as there is no pole to expose your vertical movement. In order to track your vertical motion, you would require a second pole in the room to be stationed perpendicularly to the first pole (assuming we may not use the floor as reference, since it does not represent anything in this analogy). Now, you have two reference points; a vertical pole with which you may compare your horizontal motion (walking), and a horizontal pole with which you may compare your vertical motion (jumping).

In the same manner, two paths of two perpendicularly emitted photons, regardless of the velocity or length by which the paths stretch, may be used as reference points for the absolute motion of any object. Although the perpendicular emission of two photons would certainly serve as a stationary grid in space, *tracking* such a grid may not be practical, currently. Optically seeing the path of a photon is impossible, as a photon must enter the eye to be seen. However, a method of tracking a photon's path of travel will be mentioned later on in this chapter.

Unsolved for a Hundred Years

How has the theory of Special Relativity harbored bold fallacies for over a century, without a single person coming forward to expose them? Recalling on Chapter 1's "When Fallacies Hide," some fallacies successfully sell their lies by introducing a vaguely interpreted concept and improperly applying it to a more specific situation.

For so long, no one has noticed that a vertically emitted beam of light should not travel horizontally as a result of its source's horizontal motion. One vague concept that is interpreted here, is that a "thing" will continue to move horizontally upon vertically departing a horizontally moving source. Technically, this statement is correctly used in Special Relativity's Time Dilation, as it is vaguely expressed. Light is a *thing*, and *things* do just that. When narrowing down the details, however, it is discovered that the only *things* that behave in such manners are substances that have mass. As explained, only objects with mass may accrue momentum from another object's physical movement.

The other vague concept in this scenario is that a *beam* of light will move sideways along with its source. Vaguely, this statement is correct as well. Figuratively speaking, moving a flashlight sideways will cause its "beam" to move sideways. Specifically, however, as a flashlight is moved sideways, the visible beam is not actually moving sideways. Rather, the flashlight constantly generates a new beam as the position is altered.

Nevertheless, even if people did sincerely attempt to argue that light may not move perpendicularly to its direction of travel, the argument may have been brushed aside by the notion that light's source is not *actually moving*. Although such a response may seem fair, one must realize that it comes only from the assumption that the theory is true! One may not logically argue, for example, that God is real only because "God *said* he is real." If we do not know of God's existence, then the source of such words is not credible. BEFORE considering whether or not Einstein's theory is false, one question is presented, "does absolute motion exist?" There are two possible answers to this question. One of the answers is Einstein's, "No, motion is only Relative." The other possible answer is,

"Yes, motion is absolute." We have now split our perception of a possible reality into two paths. Einstein works along his path to vouch for his claim that reality is relativistic. The other route (motion is absolute) also has statements that *should* be made. Considering the laws of momentum and the properties of a photon, the vertical path of light should *not* remain aligned with its horizontally moving source, thus exposing Jack's movement from Eve's perspective in Example 1. This perspective of absolutism never made way, as physicists were too preoccupied with testing Special Relativity's *conclusions* against its own realm of rules; inevitably surrendering absolutism. Although Relativity's conclusions seem accurate, as time variations are indeed observed in objects at varying speeds and gravitational fields, Chapter 5 offers an explanation for time variations from an objective and classical approach to physics, opposed to Einstein's relativistic exposition.

Does Light Have a Medium?

Sonar waves used by submarines depart only from the precise point from which they are emitted, using water as their medium, without experiencing a directional influence from its submarine's motion. Similarly, light waves (or photons) only depart from the precise point from which they are emitted without influence from its source's motion. This may indicate that space is the sole medium of light. Any ripple, after all, is an indication that a medium is present.

As physics will agree, light does not technically travel through anything else other than space. Water, air, or other transparent materials, are not actually mediums of light, although it may *seem* that a photon may travel through them. As a photon encounters a transparent object, it is fully absorbed by the material's electrons. It then vibrates the electrons within the object, causing *new* photons to emit from said electrons in a certain direction. This is why light's velocity appears to slow down as it *seems* to travel through objects; there is a delay between each absorbed photon and released photon. Photons only travel through space, and from electron to electron. This interaction between light and such materials, referred to as *refraction*, transfers energy, though it does not necessarily transfer *the same* photon from place to place. Thus, if there is a true medium of light, and light doesn't somehow propagate on its own terms, we may only assume that its medium is space, since it is only space through which it travels.

The Significance of Absolute Motion

Postulate 1's *first* claim is that the laws of physics do not recognize absolute motion.

There is a logical means to argue the significance of absolute motion in physics. The following is merely constructed by the deductive reasoning of perhaps the most rudimentary and fundamental principle of physics... "You can't be in two places at the same time."

MAZI'S 2nd THEORY: SPACEPOINT COHERENCE

No object may simultaneously exist in multiple places at any instant, nor may multiple objects exist within the same exact place at any instant. Respectively, no single point in space (an available space within which objects may exist) may contain multiple objects at any one time, nor may multiple points in space contain the same one part of an object

at any one time. Therefore, all points in space (or space-points), of which there are an infinite abundance, must each be regarded as individually unique entities and coordinates of space. That said, the distinction between the constant motion of an object and the motionless state of an object is indeed apparent to the laws of physics, as it determines which spacepoints are occupied by an object, and which spacepoints are not occupied by an object at any given time.

In regards to the notion that no one object may exist in any two places at one time, there is one physics experiment which has caused a commotion in the science community, leading many to believe that photons and electrons may actually travel through two places at the same time. This experiment was called the "double-slit experiment," and will be scrutinized in Chapter 7 to argue that no such contradictions may occur.

So, what does it mean for an object to be truly motionless, even if there are no other objects around it to be used as reference points? It means that the object is sitting in one point in space, without moving from one point in space, to another point in space. If each spacepoint was not a unique coordinate in the universe, then why would some spacepoints contain objects that other spacepoints do not?

Moreover, relatively to one another, spacepoints may not alter location, as they may not intersect one another. If two spacepoints were to intersect, that would mean that two objects may exist within the same location, and that two locations would somehow be present within the same location. One may then argue that "all of space may move or shift together as one unit, relatively to even an absolutely motionless Photonic Reference Point." Given that the universe is filled with an infinite amount of spacepoints, there would be no sense in assuming that they may all shift together, since they are each bordered and locked in place by surrounding space-points. For a shift to occur, *room* for shift is required, and since the universe expands infinitely, there is no "lack of space-points" anywhere that may allow a shift to occur. Thus, we may assume that the framework of space is altogether mo-tionless as is the Photonic Reference Point.

If space *did* somehow move, it would nonetheless be mo-tionless relatively to the Photonic Reference Point, since space is light's medium. Conversely, even if space is somehow *not* light's medium, the motion of space relatively to the things within it (including light's perpendicularly motionless state) would not, to our current understanding, display any known effect in physics. Space's movement (if there were such a thing) would be irrelevant to physics in the same man-ner that Einstein once believed that absolute motion would have no meaning in physics. Whether or not space is an actual entity (opposed to a lack of one), the logic of Spacepoint Co-herence yet applies. One may simply replace the meaning of "space" with "place," and Spacepoint Coherence would hold its logic. To define "space," one may say that it is a grid which is motionless together with the stationary path of any photon. The statement that space may move on its own, or *behave* as an entity apart from the objects and energies within it, simply bears no meaning.

The Space Locater

Upon consideration of the Photonic Reference Point, one way to test this theory, though we may or may not currently possess the technological means necessary to test this with enough accuracy, is by building a giant tunnel to contain a photon gun at one end, perfectly aimed to the center of a target at the other end of the tunnel. If the photon gun is perfectly accurate, but it shoots off-center, we know that the tunnel is traveling at a certain calculable speed through space, relatively to the photon's motionless path of travel. However, any velocity of the tunnel's movement through space may not be fast enough to allow the photon to hit off center, since the photon travels so incredibly fast. That is why the word "tunnel" was specified; the longer the distance the photon travels before hitting its target, the better chance we would have at detecting a shift. If this tunnel were to be stationed on Earth, we could determine Earth's true motion through space, though a second tunnel aligned perpendicularly to the first tunnel would be required to track Earth's movement in both vertical and horizontal planes. Is Earth the center of the universe (quite an egotistical assumption, but nonetheless indicated by no detection of shift of the space locater), or are we moving at tremendous speeds along with our solar system in a certain direction (more likely)?

Chapter 4:

The Ignorant Observer

A detective may ask why Special Relativity's Time Dilation (Example 1 of Chapter 2) determines that Eve's clock is ticking faster than Jack's clock. Since Eve's motion is considered to be equal to Jack's motion from each of their perspectives (that is, motionless), why would Eve's clock tick faster? Because Jack is emitting the laser? What if Eve was instead emitting the laser towards a ceiling that reflected the laser back down to her? Jack would then notice that Eve's laser is traveling diagonally, while Eve assumes her own laser to travel vertically. This would conversely result in Jack's clock ticking faster than Eve's. So, after they continue this relative motion for a year, who will have aged more? Would that depend on who was holding a laser pointer? What if they were *both* holding a laser pointer?

A detective may also ask why each observers' perception of time is based on the notion that each observer must assume oneself to be *motionless*. When the observer notices a passing object, why must the observer assume that the object is moving? What's to stop the observer from assuming that oneself is moving while the object is stationary? If, per Postulate 1, the laws of physics are the same whether you are in constant motion or motionless, then logically, shouldn't each observer be allowed to claim that they are moving? If, in Example 1, Jack and Eve each claim that they are in motion, then the conclusions of Time Dilation would be reversed; Eve would calculate a vertical path of light while Jack would assume a diagonal path of light. This would result with Eve's time ticking slower than Jack's time; opposite of the conclusions originally derived by Example 1. If their relative motion continues in this manner for a whole year before they meet again, which one of them would have aged more? Wouldn't that result depend on which state of motion they each assumed? Moreover, what physics law determines what their relative assumptions should be in regards to 'who is moving?'

In Example 1, as Eve stands on the platform and assumes that she is motionless, she times the diagonal laser to meet the ground more slowly than if it were traveling only vertically. What if, mid-way during this event, Eve changes her mind and assumes that she is in motion while Jack is stationary? Since the laser's vertical speed is slower in the case that Eve is motionless (because it is also traveling a horizontal speed and cannot exceed the speed of light), will Eve suddenly notice a change in the laser's vertical speed once she assumes that she *is* in motion? If Eve rapidly changes her mind back and forth between who she assumes is actually in motion, would the laser rapidly change its vertical speed?

What if both observers agree to have aligned assumptions? Referring to Example 1, if Jack and Eve both choose to assume that *Jack* is completely motionless, then they would each witness the laser to have traveled the same distance, eliminating the effect of Time Dilation.

Especially in the discussion of the theory of Special Relativity, physicists have unreasonably abused the fallacious concept of the supposed "observer," weighing the dependability of any course of events too heavily on the limited and ignorant perceptions of those watching the events.

Special Relativity bases its concepts on the supposition that certain aspects of the laws of physics are determined by and limited to the observer's knowledge alone. It's fair to say that we can only assume the existence of a property if we can witness it, as to prevent ridiculous claims from being accepted by the science community. Special Relativity makes a logically valid point in implying that we may not claim with certainty that motion is absolute, since we *could* not (we can now) identify movement absolutely. This was a very wise statement from Einstein, as science should not accept a concept which is supported by no evidence. Special Relativity states that we may only identify the relative motion between two or more objects; upon witnessing relative motion, we cannot be so hasty as to certainly claim that any specific one of the objects are in actual motion. We must give both objects the benefit of the doubt of being in any possible state of absolute motion, if there is such a thing. Thus far, Einstein's reasoning is perfectly modest and fair.

The absurdity begins when Special Relativity assumes, only because we cannot prove or detect absolute motion, that absolute motion *most certainly* does not exist, forcefully applying the concept to scenarios in an illogical manner. Special Relativity is assuming that because two relatively moving observers may not identify which of them are actually in motion, that both observers may *simultaneously* claim to be motionless! Why? Because the laws of physics do not recognize motion to be different from stillness, per Postulate 1 of Special Relativity. This inference is logically incorrect in three ways.

For one, logically, we know that when two observers are in relative motion, there *must* exist some actual or absolute motion, as the distance between them may not close or open otherwise. *Movement… is* occurring, surely; we just don't necessarily know where. Thus, in one given event whereby there is *certainly* movement between two relative observers, at least one of them, if not both, must be moving. For *each* of the observers to claim to be motionless within the same event where at least one of them MUST be in motion, is not only illogical, but also mathematically incorrect. The situation requires at least one mover. Thus, one may not logically derive a conclusion such as Time Dilation from the *same* event based on *two* opposing assumptions (Eve's assumption that she's motionless, contrary to Jack's assumption that he's motionless) by which *neither* assumption takes responsibility for movement.

Secondly, the observer's claims must be aligned, since they are sharing the same situation, the same event, and the same reality! Jack and Eve must *both* agree that either Jack or Eve is motionless as they decide to measure aspects of the situation. If they each have different opinions of what may be occurring, and those different opinions cause different results in their calculations, then of course they would obviously provide mixed results for the same event. That doesn't mean that their "rates of time" are different, it means that they are displaying different results from their calculations due to their difference in opinion of how to perceive the situation. Nevertheless, the extent to which results may vary between both of the observers' simultaneously opposite assumptions in Example 1 comes from nothing other than the fallacious idea that a photon may retain the same constant velocity as it erroneously accumulates horizontal momentum and accordingly a greater distance of travel.

Thirdly, given that Postulate 1 states that constant motion is the same as the motionless, then why does motion produce different results in Example 1 than does stillness? In Example 1, if Eve is not moving, there is a longer and diagonal path that Eve sees the laser travel. If Eve *is* moving, there is a shorter and vertical path traveled by light from Eve's perspective. Special Relativity claims that physics recognizes no difference between motion and stillness, yet motion produces different results than stillness in Special Relativity's very own examples! Does this not signify a difference in the laws of physics between motion and stillness? Special Relativity, along with its *own* examples which contain results of insanity, acts as a *test* to the question "does physics recognize the difference between motion and stillness?" As we answer this question with "no," all hell breaks loose as multiple and contradicting results occur in the same event.

Theories are usually rejected if they do not coherently apply to our reality. Special Relativity completely changes reality with nonsensical conclusions, and it is somehow accepted; perhaps accepted in the same manner that the hotel-riddle (from Chapter 1) would have been accepted if it were so intricate that no one would've been able to logically prove its faults, and if its supporting formulas created thereafter were so well devised that its fallacious method of accounting was mathematically adapted to successfully apply to any transaction. Einstein was a brilliant Mathematician, and so his formulas were very well devised as to allow his theories to adapt to certain matters of physics quite well. He was a mathematical computer of brute force. Whether or not his understanding of reality was skewed, though, is in serious question.

Human observers may not detect a difference between motion and stillness by using their eyes, other senses, or current technological means. Special Relativity agrees that we do not have the right to claim which certain object is moving and how much, however, it states that because of this, any object may claim to be in ANY state of motion (because motionless and constant motion are the same thing)! This statement contradicts the theory's very own logic that one may not identify motion without the *evidence* of a reference point. Since only a reference point may suggest that motion is occurring by relative comparison, then accordingly one should not identify (or assume) *its own* state of motion (as do Eve and Jack). Special Relativity begins its premise with the belief that we should not claim anything, such as true motion, to exist if we cannot witness or prove it, only to stray away from such discipline as it allows any observer to claim a motionless state without obtaining *any evidence* to do so. Moreover, Relativity claims that true motion does not exist, without holding any evidence that it does not exist.

Special Relativity could not have made its claims if it didn't utilize the aid of limited observers who are ignorant to the full truth. When we assume that Jack is in motion in Example 1, there is a *greater distance* traveled of the laser that only Eve can see. Why is Jack ignorant of this occurrence? Because he is blinded from seeing the true path and length of the photon's travel? There is indeed a difference between a photon traveling a path that is shorter, opposed to traveling a path that is longer. There is also a GEOMETRIC difference between a photon's diagonal "v-shaped" path, and a vertical "I-shaped" path. Yet, Special Relativity claims that both of those paths are the same thing, only because the incompetent observers viewing the event cannot tell the difference between them. What actually happens in reality is somehow entirely based on what the incompetent and ignorant observer *thinks* he or she sees.

Motion is thought of, by physicists, as only to occur if the ignorant and incompetent observer can *see* it occur, in which case a second object or reference point is necessary to aid the disabled observer in noticing such motion. The laws of physics should not depend on the human's limited visual senses alone, as the abilities of optical senses are limited by positions and prone to optical illusions. What we *see* happening isn't necessarily what *is* happening.

In Example 1, Eve might *think* that she sees a diagonal path of the photon's travel, only because she is not aware that she is indeed moving, considering that Jack must not be moving, since his laser remains aligned with him; a photon's horizontal motion can't be influenced by the horizontal motion of its source, thus if the laser returns to Jack, he could not have been moving. If Eve knew the correct laws of physics, she would have accurately understood the true situation that only she could be moving and that the laser traveled *only* vertically. However, she is limited by her optical senses which don't allow her to detect the true path of an object in space, and also limited by her ignorance to the aspect of physics regarding photons. No insult is intended; Eve is a wonderful person and probably a pleasant member of society. Nevertheless, neither of the observers' limited observational capabilities by any means qualify them to determine our reality.

Science must put an end to the establishment of claims that are based on the conceptual perception of the limited observer. Ironically, Special Relativity is claiming scientific knowledge based on the observers' *lack* of knowledge and ability.

The lack of means to practically prove that a single object is in motion should not conclude that an object simply *can't* be in an absolute state of motion. Is it only our knowledge that may allow the existence of the universe to occur? Are we such gods? Who would've guessed that a philosophical notion as ridiculous as "if a tree falls, and no one is there to hear it, did it make a sound?" would actually be taken to heart by the science community? If an object is moving through space, it is moving through space, whether an ignorant and partially blind observer can detect its motion or not. One is mistaken to regard motion and stillness as equals only because one cannot physically detect with their limited senses the difference between the two; there is plenty of deductive reasoning to suggest otherwise, even with the Photonic Reference Point aside.

Chapter 5:

Fallaciously Correct

It is important to note that Einstein's theories originated from conceptual ideas *first*, and developed with mathematics *second*. After initially arriving to the concept of General Relativity (his theory of gravity), he required an additional six years to conjure up the mathematics to support his idea. In such a process, one may create virtually any claim, and support it with a correlating language and set of formulas which allows the claim to coincide with reality, without contradicting any current laws of physics. Consider a simple and quick example for this… Billy may claim that only he is real, while everyone and everything around him is a simulation from within his mind. The formula to support this? The consciousness and awareness that all people *think* they possess, is an artificial intelligence unconsciously manufactured by the simulation of Billy's mind. Billy is unaware of what goes on in people's (made-up characters') minds because the process occurring in his brain which manufactures people's thoughts is automatic, just as the signals that his brain sends to his heart are automatic. Only Billy exists, he is a god, all alone in one empty universe. As *we* feel, it is actually Billy's dream that feels.

No one can really "prove" Billy's formula wrong (unless we kill him to see if we cease to exist). His formula, which is designed for the adaptation of his claim into reality, does explain how his claim may hold truth amidst our universe. This dynamic explains how Einstein's theories are difficult to refute, despite the counterintuitive logic present. To support Special Relativity, Einstein didn't only use mathematics existing *prior* to his theory; he actually invented *new* mathematical formulas which adapt the applications of his claims into physics. Give a genius mathematician six years to design his own formulas and arguments to support a theory, and he could convince science that falling trees actually don't create sound waves when observers aren't present to listen to them.

As one notices fallacies in Einstein's theory, such as that of the contradicting directions of travel of the same laser, one must not turn a blind eye to the madness only because other portions of the theory *do* correlate with reality.

Mistakenly Correct

At this point, one may either hold the belief that the theory of Special Relativity is fallacious or that this book's first two theories of physics are mistaken (Photonic Reference Point & Spacepoint Coherence). Oddly enough, there may be an alternate supposition. Have you ever executed a mathematical problem with a logically incorrect conceptual approach, and yet you attained the correct answer? Such a phenomenon may occur either by pure coincidence, or because the nature of the circumstances allows the errors to be bypassed. Consider the following example of math problems, whereby correct conclusions are derived from illogically executed mathematical approaches.

Correct Answers Derived Incorrectly:

A) $\dfrac{9}{2} - \dfrac{25}{10} = \dfrac{9-25}{2-10} = \dfrac{-16}{-8} = \mathbf{2}$ B) $\dfrac{18}{4} - \dfrac{50}{20} = \dfrac{18-50}{4-20} = \dfrac{-32}{-16} = \mathbf{2}$

C) $2^5 9^2 = \mathbf{2592}$ D) $2^5 \cdot \dfrac{25}{31} = \mathbf{25}\,\dfrac{25}{31}$

$3^4 425 = \mathbf{34425}$

$31^2 325 = \mathbf{312325}$

In problem A, two fractions of unequal denominators are illogically combined before they are subtracted. Fractions may only be combined if they share common denominators. Yet, we arrive at the correct answer. To rule out the possibility of pure coincidence based on the arrangement of random digits, we tweak the problem, multiplying each numerator and denominator by 2 before attempting to solve the problem in the same manner again. This is demonstrated in problem B. Using the same incorrect approach, we again result with the correct answer. This demonstrates that the correct answer is not a result of the manner in which specific digits interact with one another. It is instead due to the relationship between the nature of the execution (including the *incorrect* methods) and the proportional value of the numbers. This allows for the numerical values to become larger while the same incorrect execution continues to lead to the same correct result. It is in this manner that a logically incorrect theory may display correct results.

In problem C, exponential signs are ignored, and the digits are sequentially combined without any mathematical operation, and yet the results are equal to those obtained if the numbers were correctly multiplied with the exponential signs considered.

In problem D, 2 to the 5th power, multiplied by a fraction is interpreted as the digits 2 and 5 together, meaning 25, and being added to a fraction. Yet, the values of both expressions are equivalent.

As presented, there are instances in mathematics whereby an approach is fallaciously applied and may nonetheless generate correct answers. Whether this occurs by coincidence, or as a result of an underlying pattern, a fallacious concept will not *always* necessarily sabotage the conclusion.

Theory's 1 and 2 (from Chapter 3) argue that there is a definite value in any motion in reality, deeming Special Relativity's *approach* incorrect, but not necessarily its conclusion.

If Einstein's conclusion of Time Dilation is indeed valid, then perhaps his accidental discovery of this concept required a formula which *assumes* Postulates 1 and 2, while the postulates are not necessarily true in reality. In other words, Einstein's formula may *allow* each object/observer to claim that it is motionless, although that may actually be the case.

This would be conceptually similar to solving an algebraic equation (containing variables X and Y) for Y = 0, and separately solving that same equation for X = 0, in order to find the X and Y intercepts of the equation on a graph. Solving for X = 0 gives one value, and solving for Y = 0 results in another value, or point on the graph. Similarly, in Example 1 of Chapter 2, when Jack's motion is set to zero, Eve's "time" shows a certain value, and when Eve's motion is set to zero, Jack's "time" alters. As their states of motion are set to zero, they act as the "intercepts" of a larger graph, notably the "spacetime continuum" (the concept that space and time are intertwined as one function).

Ironically, Einstein's theory may inaccurately describe a true (or partially true) meaning. For instance, passing through different spacepoints at a certain rate may tamper with the velocity in which an object's particles behave at a subatomic level. Such an implication would not mean that time itself is being altered for an object in motion as Einstein claims; rather that the vibrational movement of an object's particles may slow down. Both assumptions (time slowing down vs. particles slowing down) may have physically equal characteristics, though they differ in conceptual understanding.

Time Dilation or Orbital Delay?

Einstein's theory suggests that as one's velocity approaches the speed of light, his/her time slows down. Furthermore, traveling *at* the speed of light would completely *pause* one's time. If Mazi's theories are correct, and Einstein too is somehow correct, then we may assume that although the passage of time does not change, one enters a *physical* state of slow-motion as they approach the speed of light. Though, by what means may this mechanical influence on one's subatomic structure occur while simply traveling through points in space at higher velocities? To answer this, we ask two simple questions.

One: "What is the fastest possible speed of travel allowed by the universe?" That would be, the speed of light.

Two: "What movement occurs in atoms?" In any atomic structure, the main movement occurs in the electron, as it orbits the nucleus of an atom, occasionally bonding to other nuclei.

Bearing in mind that absolute motion *does* exist, excluding Postulates 1 and 2, suppose the following scenario...

An atom (containing a neutron, a proton and an orbiting electron) is traveling in some *forward* direction at the speed of light. Given that the speed of light is the *fastest* possible speed of travel through points in space, the electron may not orbit the nucleus in any direction that is *forward*. Given that the atom is moving forward at the speed of light, if the electron was to move at all forward relatively to its nucleus, it would be traveling through points in space at a rate that is *faster* than the speed of light, which according to physics cannot happen. Keep in mind that we are not assuming Special Relativity's rules; this scenario assumes that absolute motion *does* exist. This atom is not considered to be motionless while traveling the speed of light relatively to some *observer*. This atom is in true motion—traveling relatively to points in space.

So, the only direction in which the electron may orbit its nucleus in this case, is vertically from side to side. It may orbit either clockwise or counter-clockwise in this manner, though it may not move by *any* increment forward.

Given that an electron's orbital movement may vary in direction, the electron may occasionally move slightly backward relatively to the atom's direction of travel. Each time that the electron moves the tiniest bit backward, it gets stuck further and further back, until it is directly behind the nucleus, not to move at all.

Since the electron orbital movement is very rapid, it will assume the stationary orbital position directly behind the nucleus very quickly upon the moment that its nucleus attains the speed of light. Thus, when electron orbit halts, the motion occurring within the atom is "paused." As the atom's speed of travel slightly decreases, its electron may then slowly move forward; its orbital speed is slightly increased, though still limited.

As a person travels *forward* at the speed of light, not only would all of his/her electrons halt orbital movement; moreover, none of the person's constituents (atoms, molecules) may move in any direction forward *relatively to one another*. This would result in limited molecular movement of the entire body.

While considering the possibility that absolute motion does exist, the rate by which objects pass through points in space may influence the behaviors and velocities of their subatomic particles; that is, if the universe indeed reinforces a speed limit. Considering the concept of absolute motion, one may see how Einstein's Special Relativity somewhat translates over to a correct conclusion, though with a fallacious understanding of reality. Special Relativity *assumes* one's motion to be zero, relatively to the speed of light, to infer that its time slows down as it approaches the speed of light (a slow-motion effect). In both Mazi's and Einstein's theories, the molecular movement of an object may slow down as its total speed *through space* approaches the speed of light. Whether you believe Einstein in that motion is only relative, or Mazi in that there is absolute motion, you will yet retain identical physical results at speeds relative to light.

General Relativity

Sidetracking for a brief moment, it is worth mentioning an aspect of Einstein's other theory, General Relativity, which pertains to his description of gravity. In it, he describes that gravity also alters time by bending space. Thinking from Mazi's absolutive perspective, the force of gravity may perhaps slow down the orbit of an electron in the direction *away* from the gravitational source. Moreover, relatively to one another, all movement of the molecules and atoms of an object may slow down in the direction away from the gravitational source. Thus, as gravity is increased, an object's particles may slow down (given that the object were to remain in the same position, without moving towards the gravitational source). This, too, is a manner by which one of Einstein's theories could have been fallaciously correct. In relative terms, he speaks of *time* slowing down. In Mazi's absolutive terms, the particles of an object *mechanically* slow down.

One may then ask if there is a difference between the slowing down of time, and the slowing down of particles. If so, what aspects of reality would draw this distinction?

Chapter 6:

Absolute Time

Suppose that the entire universe suddenly slows down. As an electron revolves around a nucleus at a small fraction of the usual speed, is time slower for that particle, or is the particle mechanically influenced? One may define time as a "passing of events," with which the argument may be asserted that if no event occurs, time may not exist. Basing all of reality on the limited senses of an observer alone; even *time* could only exist if two separate events chronologically took place for contrast. Moreover, if time were to slow down or speed up, no observer may notice, as the observer's interpretation is equally adjusted to the rate of time. To define the true meaning of time, a particular concept of an instant must be understood.

The Universal Instant

Suppose that you somehow instantly pause the whole universe with the snap of your finger. Everything stops, including photons. Now, we are not taking into consideration different time zones. If you snapped your finger at 3:00pm eastern time, that does not mean that those residing in pacific time will freeze 3 hours later. Neither does it mean that the pausing effects of your finger may only extend outward at the speed of light, as that would result in a delayed pausing of distant objects. For this scenario, as you snap your finger, the objects around you pause, the objects surrounding those objects pause, and so forth to the end of the universe, without any delay in between your finger snapping and the edge of the universe pausing.

As everything in existence is simultaneously and instantly paused, you then project a 3D image of the entire universe into a computer. As you scroll through this 3D image, you would see a clear-cut picture of the location of all objects and energies relatively to one another at that given instant. We may call this projected image of one instant a "universal instant," which one may think of as a "snapshot" of the universe. Now, suppose that you snap your finger again to allow the universe to move, and then you snap it once more to pause it again. You project yet another image of the entire universe. You continue to do this, collecting many evenly spaced frames until you have charted a motion-picture. If you play this video, you will find that you have one constant clock running which governs the entire universe.

Observing a video constructed in this manner, regardless of how fast objects are moving relatively to one another or relatively to the speed of light, there is yet only one universal clock which oversees all activities per any given moment. There is no subjective idea of a time that is dependent on the relative speed or distance between objects, as Special Relativity attempts to suggest. Furthermore, two contradicting things (such as Jack and Eve noticing a different path of light)

68

may not occur at the same time within any one event, given that we have only one video, and only one frame per any instant. That said, having Spacepoint Coherence and the Photonic Reference Point in mind, we may deduce the theory of Absolute Time.

MAZI'S 3ʳᵈ THEORY: ABSOLUTE TIME

Each spacepoint is bordered by other spacepoints, as they allow for information (particles) to be transferred between one another based on the set of formulas (characteristics) by which the information may behave as it is transferred or stored. The interaction between spacepoints are synchronized in such a manner that they do not allow for contradictions to occur, such as multiple pieces of information being present in one location or one piece of information being present in multiple locations within the same universal instant. (Universal Instant: The relative position of all objects and energies in existence at one given moment). This limited manner by which space is synchronized, is time. Time is independent of the presence of objects, and only dependent on the formulative synchronization of space whereby particles or energies "may" be stored or transferred in a non-contradictive manner between any two given universal instants. A change may not occur without time; therefore, time is the allowance of change of information to occur in a manner limited by the synchronization of space. Time allows an infinite number of increments of chronological change in space. As time continues to allow such change in space, the universal clock continues to run, whether change is occurring or not. Conclusively, time is space's synchronized sequential allowance of non-contradictive change to occur.

In other words, individual universal instants are connected in a chronological order whereby the next frame is bound to and dependent on its preceding frame's state. This *chronological* order is a formulative sequence which ensures that each universal instant is connected without allowing unaccounted/contradictory change (transfer of information) to occur. Another example of unaccounted/contradictory change, is for an object to move from one place to another by *skipping* some of the distance in between. All actions must have connected cause. If the universe is a program, then the space-points are storages of memory, and time is the formula which *allows* for information to be adjusted within the memory storage. Time may continue to exist, even if there are no two events to be used as points of reference. As long as events are *allowed* to occur, time goes on.

We may deduce that an absolute and universal time may exist relatively to the unchanging speed of light through space. If *time* were to change, so too would the speed of light, together with any observer's perception. Thus, an alteration of time is undetectable, and the possibility of its occurrence is not determined; nor is its relevance to physics.

Slow Time vs. Slow-Motion

Einstein suggests that time is different for relative observers as their velocities vary, even within the same event. In Example 1, Jack's *time* is supposedly different from Eve's time, although they are coexisting in the same universe. This definition of time would lose meaning if it were held comparative to the frequency by which individual universal instants pass. As an accumulation of universal instants are observed, objects within this video moving at varying speeds relatively to one another (say, Jack and Eve) may show that their subatomic particles move at different rates, implying "slow motion" as

Special Relativity suggests. However, they would yet be held accountable to the same universal clock which tracks the exact location of every existing object relatively to one another, at every given moment (or universal instant).

Herein lies the difference between Einstein's concept of altered time (within objects moving at differing velocities), and Mazi's inference of absolute time. An absolute time accounts for the relative position of all objects at any given universal instant. When Special Relativity states that "time slows down for an object that approaches the speed of light," Mazi would instead state that "the object's particles slow down at a subatomic level, while time remains constant for that object and equal to the time of all surrounding objects within the same universe."

Chapter 7:

The Double-Slit Riddle

In the discussion of absolute motion and space, two concepts of absolutism were theorized in attempt to challenge Einstein's concept of a relativistic reality. The Photonic Reference Point was met by reference to the known characteristics of the photon, elucidating that it holds a motionless state perpendicularly to its direction of travel. Secondly, Spacepoint Coherence deduced that the motion of an object must involve a change of location from one spacepoint to another, implying the existence of absolute motion.

A particular experiment in quantum physics suggests that a photon or an electron may possibly travel through two places at the same time. Even more, it is concluded that they may exist as two different things (objects and waves) at the same time. Physicists have dubbed this phenomenon the "wave-particle duality." If such a concept is true, then the idea of Spacepoint Coherence may omit to such exceptions in the small-scale behaviors of subatomic particles. Nevertheless, it is worth examining the process that led physicists to believe in the wave-particle duality. Perhaps, as the idea seems immensely unorthodox and counterintuitive, it may have been concluded with faulty reasoning.

The Double-Slit Experiment

The "double-slit experiment" was first performed by Thomas Young in 1801. In the basic version of this experiment today, a laser beam is emitted onto a plate with two narrow, parallel and vertical slits (holes which look like "I I"). As the laser passes through the slits, its luminosity is observed on a screen behind the plate.

When light (an electromagnetic wave, made up of photons) is emitted onto the plate, the single wave of light is blocked by the plate. Only two portions of that wave are able to exit through the slits (holes) in the plate and toward the back screen. As the two separated narrow waves exit their slits, they each expand and overlap one another as they move toward the screen. When the two waves overlap, before meeting the screen, the waves interfere with one another. As this wave interference occurs, sections of the waves destruct (disappear) while some amplify (get brighter) and others remain the same. The effects of the wave interference are displayed on the screen as bright and dark vertical bands. The bright bands are areas of the screen on which there is a presence of light, and the dark bands are areas which lack light (indicating the destruction of light waves). This pattern of bands is referred to as a wave interference pattern. Figure 3 on the following page illustrates this experiment from an overhead perspective.

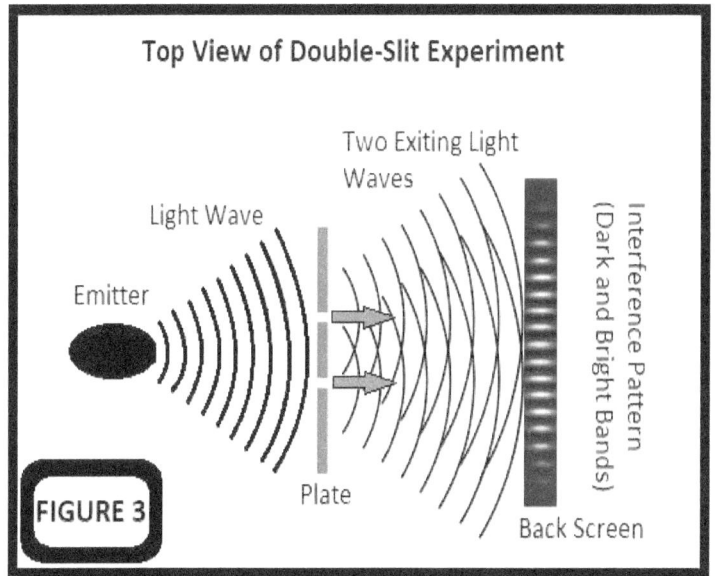

Top View of Double-Slit Experiment

Two Exiting Light Waves

Light Wave

Emitter

Plate

Interference Pattern (Dark and Bright Bands)

Back Screen

FIGURE 3

Waves of all sorts interact in the same manner, be it radio waves, sound waves or even water surface waves. The pattern displayed on the screen of the double-slit experiment infers that light is a wave, since wave interference patterns are only caused by waves.

In more recent times, as physicists began to fire single photons (particles of light) through any one of the two slits, single dots would appear on the back screen, implying that light is made of singular, tiny particles. Thus far, nothing is unordinary, as light may be thought of as waves which consist of many tiny photon particles.

The following is what dazzled physicists, leaving them mystified to this day. If many consecutive photons are emitted through the slits only one at a time (regardless of how much time passes between each emission), the dots on the backscreen gradually accumulate to form the same wave interference pattern that a stream of light would produce.

This does not make even the slightest bit of sense to scientists today. If a single particle of light is traveling alone, there is no second wave for it to interact with, thus no wave interference pattern should occur. Yet, somehow, each photon "knows" the precise locations at which the previously and subsequently emitted photons meet the backscreen as the photons cooperate with one another (again, fired at separate times) to form a pattern which suggests that two simultaneously emitted waves are overlapping.

To add on to the confusion, further bewildering physicists, closing one of the two slits as they continue to fire single photons (one at a time) through the open slit, eliminates the interference pattern formed by the accumulation of separately fired photons, evenly filling the entire screen with dots.

The question is, how does an interference pattern form on the screen by accumulation of separately emitted single photons, ONLY when both slits are open? How do the photons *know* when both slits are open?

The perplexity brought onto physicists by this conundrum led them to conclude that photons are neither particles nor waves, but rather particles *and* waves concurrently, given that one photon particle may exhibit wave-like behaviors. They have also concluded that a single photon somehow travels through *both* slits simultaneously, to interfere with *itself*. This concept is dubbed the "wave-particle duality." Electron particles also possess wave-like characteristics, as they too display wave interference patterns when fired through slits in the double-slit experiment.

Solving the Double-Slit Riddle

It is contradictory to all that science has ever observed before the double-slit experiment to assume that one thing may be doing two different and contradicting things at the same time. It would be more reasonable to construct an assumptive description of a photon that is coherent with all that we have observed of photons, including this experiment, while abiding by reality's limitations that we have honored for good reasons thus far. One mere experiment, of the millions that humanity has ever conducted, is the only one to suggest that one thing may be two completely different things at the same time. Although it is possible that a brand-new property amidst us is discovered, that should be the *last* assumption after all other possibilities are imagined. One must treat mystifying results as the riddles they are, interrogating every aspect of the problem for clues in a strictly logical manner.

Before attempting to solve the real-life riddle presented by the double-slit experiment, the key elements and properties involved must be listed, identified and understood; for they may hold hints. The experiment included one electromagnetic wave (light, which consists of photons), a plate which blocks the electromagnetic wave, two parallel vertical slits (holes) in the plate which allow two small portions of the one electromagnetic wave to pass through the plate, and a back screen which is met by the two waves. We also have electrons, which are present in the identical experiment by which electrons were individually fired instead of photons.

Photons are massless particles, and so they travel at what is referred to by Einstein as the fastest speed possible within the universe, about 300,000 kilometers per second. Photons travel as waves, as suggested by the double-slit experiment.

Electrons are particles that do have mass, about 1/1000th the mass of a proton (not to be confused with "photon", as they are similarly spelled). Electrons, having mass, are known to travel only a fraction of the speed of light.

What is an electromagnetic wave, and how is it produced? Each electron outwardly extends negatively charged electric field lines. For clarification, electric field lines are lines that are drawn outward from electrons and inward toward protons, as they represent the electrostatic force between negatively and positively charged particles. Electric field lines are believed to extend outward from electrons at the speed of light. When an electron changes speed or direction, the electric field lines extending outward from the electrons thereafter display a diagonal shift, relatively to the electric field lines distributed before the electron's change of speed or direction occurred. This, in turn, is a wave which accordingly travels at the speed of light. This ripple, or wave, is the electromagnetic wave; produced only by an electron in the act of changing speed or direction. This bond that electrons have with each surrounding particle in the universe, experiences ripples as electrons change directions, much like a rope that is held between you and your partner would experience ripples. Thus, a spherical wave is emitted in all directions, as a ripple is created in the bonds between an electron and all surrounding particles. Do note that these bonds may not necessarily be of electric field lines; they could be bonds of gravitation or possibly even quantum entanglement. However, because a ripple is present, the existence of some sort of medium or bond is suggested.

So, we know that light is the fastest particle in the universe because it has no mass, and that an electron also travels as a wave, but is slower than light. We also know that both photons and electrons project singular dots on the back screen of the double-slit experiment, and after many dots are accumulated, they mysteriously create a wave interference pattern. Lastly, we know that this interference pattern only occurs when both slits are open, somehow implying that each slit allows one wave to pass through simultaneously, despite the fact that only one photon is emitted through only one of the slits at a time.

Prior to reading the following theory, you may consider the mentioned clues and attempt to generate your very own theory that explains the results of the double-slit experiment!

MAZI'S 4th THEORY: WAVE-PARTICLE DISTINCTION

Given that each photon emission causes no more than one dot on the screen, we must state the fact that there is only one photon particle involved with each emission. Separately emitted photons exclusively meet the screen at which points a wave interference pattern in this scenario would meet the screen. Given this fact, we must conclude that two waves are indeed exiting the slits simultaneously, since the displayed interference pattern requires the interference of two waves which exit the two slits. The passing of the two waves through the open slits suggests that one larger wave met the plate before separating through the slits.

Thus far, we know that each emission involves one emitted photon and one emitted wave towards the plate (whether or not they are the same thing). We also know that one photon and two waves exit from the open slits from the back of the plate. We know that the photon may only pass through one slit, since one thing may not exist in two places simultaneously. Thus, from each single photon emission, we know that each slit allows one wave to exit, and that only one of the two slits allows one photon to exit. These facts suggest that the photon and the wave emitted are two separate entities.

79

Since the two waves exiting the slits (after the single wave is divided by the plate) do not imprint a wave pattern on the screen, we may conclude that the wave(s) are invisible and undetectable by the screen. Considering that only the photon is visible (imprints the screen), and that each emission allows a photon to meet the wall only at which points the interference pattern suggests that photons may be struck, we may logically assume that a photon is traveling in a manner suggested by the invisible waves.

To comprehend this proposition with ease, one may visualize that on some given point within the invisible wave (before the plate divides the wave into two), a single photon particle is permanently stationed, as if "surfing" on that certain point of the wave. As said point of the wave interferes with other waves, the wave alters, accordingly altering the location of the photon. Where the wave ceases to exist due to wave interference, the photon cannot surf due to a lack of wave. Consequently, each emitted photon may only meet the screen at which points are met by the invisible wave. As many emissions occur at separate times, each of the photons surf on certain points of their own waves; the position of each photon relatively to its wave will vary from photon to photon, as the specific subatomic location from which point the photon is emitted varies. Each photon may only meet the screen at which points the waves tend to make contact. Each emitted invisible wave is identical to all previous and following waves, as waves are emitted from their source spherically in all directions, and both the location of the source and the slits remain unaltered; therefore, each wave will meet the screen with the same identical interference pattern. Conclusively, electromagnetic waves and photons are two distinct entities. When a single photon is emitted, an invisible electromagnetic wave is distributed in all directions, within which the photon surfs on a certain point of that wave. Since electromagnetic waves transport photons, and electrons absorb photons, then it follows that electromagnetic waves may also move electrons, hence the results of fired electrons in the double-slit experiment.

Wave-Particle Distinction

If photons and electromagnetic waves exist as two different *things*, what happens to a photon surfing on the point of a wave that is destroyed by wave interference? A photon is energy, and energy cannot be created nor destroyed, rather only transferred. When wave interference patterns occur on the surface of water, waves are not actually 'destroyed.' That would have to mean that the water particles forming the wave are also destroyed. Through wave interference, waves are simply distorted and moved around, causing a concentration of waves in some areas and an absence of waves in other areas. Thus, the dark bands on the backscreen where there is no presence of light simply indicates that the waves have moved to other locations, along with their carried photons, resulting in brighter bands to exist in other areas with higher concentrations of photons.

As an electromagnetic wave is emitted from an electron, energy lost from the electron may transpire into the outgoing wave as a photon. Given that photons do not have mass, the idea that they are physical particles should not be accepted, logically. More reasonably, photons may rather be perceived only as energy that responds to (or is part of) electromagnetic waves. When an electron is struck by a photon, its energy level increases. It then moves more rapidly around the nucleus of an atom, extending further out and away from the nucleus.

Conversely, it is known that a photon (and electromagnetic waves) is emitted from an electron when an electron loses energy.

If an electron is a particle that has mass, and a photon is massless energy, we may logically assume that an electron is composed of photonic energy which may be absorbed and/or released by the electron. We may say that while the motion of photonic energy may be influenced by electromagnetic waves, an electron is a particle which is composed of photonic energy; and so, it's motion may also be influenced by electromagnetic waves. Evidence for this is the speed at which electrons and photons travel. If photons, as well as electrons (due to their photonic energy), are transported by (or surf on) electromagnetic waves, the difference in mass between electrons and photons would explain their difference in speed. Given that photons are weightless, they should match the speed of the electromagnetic wave moving them (and they do, indeed). If an electron was to be moved by an electromagnetic wave, it logically should *not* match the speed of the wave, considering that it has a tiny bit of mass with which to resist some of the electromagnetic wave's force; thus, deeming electrons slower than photons. The motion of electrons caused by a passing electromagnetic wave may only be vibrational if the electrons are firmly bounded to an atom—they will budge, but they will hold their ground. A loose electron, however, will travel along with an electromagnetic wave at great speeds, but again, not at the speed of light since it has weight, unlike a photon.

For an electromagnetic wave to be produced, an electron must change its speed or direction, causing a ripple to form in its electric field lines which extend outward from the electron. As the electron causes this ripple, it also loses energy (photons) which is transferred via the ripple. The concentration of photons that are present in a certain type of electromagnetic wave may perhaps determine the wave's visibility, while the frequency of electromagnetic waves determines their colors. It may be that an electron *always* produces ripples in its electric field lines (thus electromagnetic waves) since it is always moving; only surrendering a photon to the outgoing ripple if the electron's change in movement was sufficiently forceful.

The Observer's Intrusion

Thus far, Mazi's wave-particle distinction explains the events occurring in the double-slit experiment without altering the classical concepts of distinct "particles" and "waves." There is, however, one more unmentioned issue presented by the double-slit experiment.

As physicists pondered whether or not the single-fired (one at a time) photons may have been passing through both slits simultaneously to interfere with *themselves*, they wondered if they could use a device to detect which slit(s) precisely the photons were passing through. The moment they turn on the device, the photons cease to develop accumulated interference patterns on the screen, and only pass through one slit at a time. The scientists' conclusion? "The photons know when they are being watched, and so they stop being *naughty* only when we watch them." These conclusions are coming from educated physicists. Although they are great at mathematics and recording data, they are not exactly the Aristotles of our time when it comes to deep thinking.

By what means are physicists "watching" these particles pass through the slits? Not with their eyeballs. They are using electronic devices and signals to detect the photons as they pass through. Electronic devices... emit electromagnetic waves. They may even create magnetic fields. If such devices may "detect" photons or electrons, then they are surely *interacting* with them in some way. Is it coincidence, then, that as they turn off this device, it no longer interrupts the experiment? Only when the device is on, may it emit electromagnetic waves or magnetic fields. Thus, only when it is on, may it interrupt the experiment. The precise manner in which it does conflict with the experiment is the only mystery here; not whether or not particles "play with themselves only when we're not watching."

Physicists are introducing electronic machines, which emit electromagnetic waves and magnetic fields, to the close proximity of an experiment which is entirely based on the study of electromagnetic waves. They should not be surprised that a disturbance, or change, may occur. Although the physicists performing such experiments think that they are merely "observing," inferring that their observations alone altered the experiment's results is a faulty conclusion, for they are bringing unknown factors into play when they introduce devices to the environment. It is not only the human awareness that is introduced in such a case.

If a photon truly behaves differently in response to it being *observed*, then logically it would discontinue its interference pattern as the *interference pattern* is being observed. The interference pattern is just as much a product of the photon as is the direct detection of a photon. There is no logical reason to assume that one type of observation would affect the experimental results differently from another perspective of observation. The only difference, in this case, between the two methods of observation is whether or not an electrical detection device is involved. When it is involved, *then* the experiment displays different results. Thus, the only element causing changes is said device, and *only* when the device is on.

An Objective Reality

Mazi's theory of wave-particle distinction coherently aligns with the mentioned general behaviors of electrons and photons, including the mysterious behaviors observed in the double-slit experiment. To date, no other reasonable answer has been theorized for the dazzling puzzle provided by the double-slit experiment. Although this answer could be incorrect, it is far more reasonable than jumping to the conclusion that particles are being naughty; doing things they shouldn't be doing when we aren't watching.

The logical deductions thus far were demonstrated in hopes of conveying the notion that all things in life, even the most seemingly ambiguous, may be objectively understood as long as the correct logical scrutiny of all relevant concepts is feasible.

There is no rational place for assumptions such as "one thing may be two completely different and contradicting things at the same time" (wave-particle duality), or "one thing is doing two contradicting things at the same time" (one photon displaying two separate paths simultaneously, as per Example 1's Time Dilation scenario). Not unless, of course, a just explanation is discovered. Evidence has *reasonably* shown science thus far that there may exist a definite description for any one entity, and it may at times be narrowed down by a logical deduction of the analyzed concepts involved.

Herein lies a bold claim that there is no truly random, contradicting or indefinite property in reality. Though, a displeasing notion is indelicately implied that humans may not possess free-will. If every action, behavior or thought is constructed by a definite process of physical nature, then how may humans retain any sense of worth or meaning in a reality so drearily predetermined?

Chapter 8:

Determinism vs. Free-Will

As trillions of microscopic particles of dust swirl with the wind in every which direction, it is claimed that the exact shape, texture, position and movement of each particle per any given moment was forever predetermined. Say, a particle of dust meets the corner of your left eye at exactly 5:32:06 PM on February 3rd of 2026. This exact event, including the specific dust-particle involved, is believed to have been prearranged to occur ever since the beginning of time. If there *is* no beginning of time, then it is assumed that it has forever been determined.

This is not a mythological notion of fate or destiny, but rather a mathematical concept referred to as "determinism." It claims that no event in the universe is random; each event occurs in a precisely direct response to its preceding situation. In other words, any event in question has a cause from which it was orchestrated to the finest detail. The contributing factors responsible for the specific results of any subsequent event include (but are not necessarily limited to) the position and state (state of motion, energy, charge etc.) of every subatomic particle, or any other force that may or may not be known to us, present within and surrounding the event. Although such depths of any situation may perhaps never be fully calculated by any human or technological device (due to the unspeakable number of moving particles and variances involved) events are predetermined by their preexisting situation's state which holds only one inevitable future.

This concept applies to all events, including social interactions. According to determinism, every decision or thought that one produces is precisely caused by (but not necessarily limited to) the physical, chemical and neurological state of one's brain as it is influenced by its stimulating environment, genetic makeup, and past experiences.

Many people, even some scientists, disagree with this concept, as they refuse to accept that the course of life may be so automatically fixed. Furthermore, this concept is widely perceived as an implication that humans do not possess free-will, as many of us intuitively believe we do. The term "free-will" is described as the ability to act at one's own discretion, without the constraint of fate. The concept of determinism suggests that we do not have free-will because our actions and thoughts are purely an automatic result of our previous situation(s), hence it is not we who control our fate, but rather the particles from which we are made and their automatically responsive mechanisms. The debate between believers of free-will and determinism has been ongoing, as it lacks an indisputable argument from either side.

As applied in preceding chapters, the concepts and meanings within the topic will be scrutinized in attempt to reach one logical and absolute conclusion for this debate. One may find that logical issues are present within *both* sides of this dispute.

On the free-will side, it is believed that there is no definite fate. The argument to support this is that life is truly random and incalculable. Contrarily, determinism believes that nothing is random; if we were hypothetically given the ability to limitlessly analyze, calculate and understand the functions of the infinitely smallest constituents of all matter and energy, we would be able to explain and make sense of any event, as well as accurately predict any event before it occurs. Although humans do not possess the resources to calculate or inspect the universe with such detail, determinism believes that all events nonetheless occur in a precise cause-and-effect manner.

The fulcrum of this debate is the term "random," as its identity is assumed to determine the winner of the argument. If life is truly random, then events may not be predetermined by previous events. On the other hand, if the quality of randomness does not exist, then all events are strictly determined by a cause-and-effect nature. The term must be scrutinized to determine whether something may be random, not random, or random to a degree.

True-Randomness vs. Cause-and-Effect

For something to be random, it must be unpredictable. What does it mean to predict? It is the act of utilizing the information that describes the characteristics of a given scenario to determine what event will follow. If a boy's hand releases a tennis ball, the predictor of the subsequent event must first consider the information present within the situation. The boy is on Earth, so there is gravity. Gravity pulls down on objects such as this ball. The ball is bouncy, and so the height from which it is released will cause the ball to fall to the ground and to bounce back up. The direction of the ball, and where it may land, is considered very predictable by a typical human since there are not many factors in play. How high the ball will bounce after falling and how many times it will bounce, however, would prove more difficult for any average person to accurately predict. A physicist, though, who has acquired detailed information such as the weight, size, density, and flexibility of the ball, along with the material property of both the ground and the ball, may very accurately predict the course of events to follow (granted there is ample time to construct the necessary calculations). The predictability of anything depends on the predictor's ability to predict it; it is a relative term between the predictor and an event.

If the boy's ball was actually a helium filled balloon disguised as a tennis ball, the physicist's prediction would be incorrect if he did not possess this true information. The ball would float upwards, and the physicist would be momentarily surprised. So, for a prediction to be accurate, sufficient information must be considered.

If one were to drop a handful of marbles onto a table, one may say that the marbles will bounce in *random* directions. In such a scenario, the term "random" is used because no one person may predict in which direction each marble will bounce after contacting the floor and colliding with one another. Suppose, however, that a physicist is to view a three-dimensional recording of this event. The video ends just before the marbles make contact with the floor. Given that the video allows the physicist to view the event from any angle that he pleases, if he were to inspect the video frame-by-frame, analyzing the motion of each marble before it makes contact with the ground, he may very well predict how and where each marble will end up moving. To an adequate physicist with access to sufficient information, the marbles will not move in any *random* direction; they will move in specific directions. From this, we find that the term "random" is only relative between an event and the observer attempting to understand or predict the event.

As Einstein argued that motion only exists relatively to reference points, the term "random" may only exist relatively between the predictor and the event in question. Although Mazi's 1st theory (Photonic Reference Point) may argue that absolute motion exists, even an object's absolute motion exists relatively to the space through which it passes, or to photonic reference points. Nothing "happens" without a relative element which offers a *description* of the occurrence.

On the other hand, the statement that a thing may have a quality of *absolute* randomness, irrelevant of any observer trying to predict it, just has no meaning. To explain that something is generally random only states that it is unpredictable by the general observer. The movement of a handful of bouncing marbles are random to the general person, but not random to an adequate physicist utilizing sufficient methods of prediction. Even to say that an event is *completely* random only means that there is no one observer that may predict the event due to a lack of resources or knowledge; it does not mean that there *is* no direct cause for the event.

May an outcome be considered random if the mechanisms of its cause are fully understood? If the cause is fully understood, then the results too may be understood. Stating that an occurrence may be *truly* random, disregarding observers, implies that nature simply "poofed" the event into existence. If nature does poof something into existence, though, then by what means and in what manner did nature poof said thing into existence? If something *happens*, it happens in a certain way, otherwise it may not be claimed that it happened at all. Something cannot "kind of" happen, or "partially" happen. It either happens, or it doesn't.

Beware of a play of words that may confuse the meaning of "something happening." One may not logically say that "a glass being filled" may *partially* happen. If the glass was partially filled, then it was indeed *not* filled, thus the event of "a glass being filled" did NOT occur. Instead, the specific event whereby "the glass is partially filled" occurred. Any event, as long as it is specifically described (to avoid confusion from a play of words), may either occur or not; there is no third option of "kind of occurring."

Thus, if nature "poofs" something into existence, the "poofing" *happened*. If something *happens*, then the occurrence exists as an event. Anything that exists, holds a definite description or property, as it may not partially exist. In conclusion, if the universe poofs something into existence, the poofing occurs in a certain and definite *way*, meaning that it came about by some means and was not random. And so, even an event being poofed into existence is not truly of random nature—something must have caused the poofing to occur.

To simply understand both sides of the argument regarding true-randomness, an analogy may be considered. Suppose a straight row of dominoes are set from a start-line to a finish-line, analogizing the cause-and-effect nature by which events occur in life. Prior to tipping over the first domino, most would agree that the following course of events will not be random. Everyone may expect each domino to tip over the next, until the last piece falls over to the finish line. This analogy describes a non-random reality, as each event is directly caused by the previous.

Now, suppose that the same row of dominoes is set, but this time there is a large gap in the middle of the row. Most would expect the second half of the row to remain standing after the first half is toppled over, given that the falling domino before the gap may not reach the next domino after the gap. As an analogy, this design suggests that the gap is equivalent to a missing link in between any two events, after which events will cease to occur (given no more connected cause). The concept of true randomness, however, would allow the domino after the gap to fall just after the previous domino falls, even as no contact is made between the two pieces. True-randomness claims that, although a link between any two events may be missing, the subsequent event would somehow occur by its *own means*, without a direct cause to result in its behavior. (Occurring by its "own means" is no different from an event "poofing" into existence, as discussed previously; which, again, would not be random.)

The issue with such logic is that, if a domino may fall by no cause whatsoever, one must also accept that the domino may do *anything*. Why fall? Why not jump up and down? Why not teleport into your mouth? Each one of these scenarios hold equal logic to the notion that it may fall without cause; if it may fall without cause, then it may do *anything* without cause.

Specific events, such as falling, only occur as a specific response to a situation. A domino falls because it is pushed forward, off balance, as gravity tips its top-edge over. On the contrary, an element of randomness should cause *anything* to occur, since no specific instruction is implemented for an action. If there were truly any missing links between any two events, and yet the second event somehow proceeds by a truly random nature, our reality would experience constant and inconsistent chaotic change. One might then say that things may be *kind of* random, only to the degree that our reality is not chaos. The fault with such logic is that the term wouldn't hold any meaning. There is either a link in between two events, or there is not. The link cannot be *vague*; it cannot *sort of* cause something, as described with the "filling the glass" example. Something either happens or it doesn't. Humanity has yet to think of a way for something to *kind of* occur.

Even if the link in between two events were governed by some other dimension that we may not observe, to say that the domino after the gap was toppled by a ghost for instance, the *governed action* is yet something that is *happening*. When something is occurring, it is occurring in a certain way, and it is thus definite. If a ghost, which we may not observe, throws an object, the ghost does so in a certain way, and so the action is directly caused. The term "truly-random" simply holds no meaning; it is a play of words.

True-Randomness vs. Timelines

Another concept to consider as one attempts to determine whether or not something may be truly random, is the presence of a timeline. As long as a timeline exists in a reality, regardless of how many time lines there may be, there will always be only one inevitable arrangement of outcomes for each single timeline.

A single storyline (or timeline) has only one line of events. This means that only one story will play out. When the story is over, regardless of how many times we look back through history to read the story, it will forever be the same one story. There will never be a different first president of the United States, other than George Washington. The twin towers will never have collapsed on a date other than September, 11th.

Thus, since there is only one set of events for any one time, there is only one definite future. A *single* timeline will not have several, constantly altering futures. Given one storyline, all events within the storyline are set and predetermined. The only means by which a future may not be predetermined (or set), is for a single timeline to have the ability to constantly change. In such a reality, every single day, we learn that there was a different first president of the United States. On some days, there never was an America to begin with. And on some days, September 11th never happened. Each day that we wake may fall on a random year... This is a *random* timeline (But again, this randomization would be governed in some definite way, which nevertheless deems it not random). Given one time line, we have but one past, one present and one future. Unless we can alter the manner in which a timeline functions, we must accept that we have only ONE inevitable ending. What it will be, *we* cannot know, because it is unpredictable by us, therefore random to *us* as observers. A future "being random" to anything other than an observer's attempt to pre-dict it is meaningless. Regardless from which point in time an event is observed or awaited, it will nonetheless exist in a de-fined way in the specific time of its occurrence; its present is experiencing it, its future is aware of it, while its past may be cluelessly awaiting it.

Note that as the theory of multiverse argues that each timeline may branch off into multiple (or infinite) timelines, said timelines would nonetheless be constructed by some def-inite means.

Therefore, any reality containing a timeline-format (even heaven, if it exists) will have a concurring element of a fixed arrangement of events.

The Irrelevancy of True-Randomness

Bypassing all logic, let us consider for a moment that this idea of absolute-randomness is true. Would the concept of true-randomness actually support the idea of free-will? Why must molecular mechanisms contain true unpredictability in order for humans to possess free-will? If free-will is the freedom to decide and act at one's own will, then wouldn't an element of true-randomness within one's cognitive process actually *take away* from one's control over his own decisions? If our thoughts are resulting from some truly random and undefined flow of events, then we are not creating the mentioned thoughts ourselves; the randomizing process is. Thus, we cannot claim responsibility for random thoughts any more than we can claim responsibility for predictable and predetermined thoughts. Whether our thoughts are a product of a random nature, or a predetermined nature, they are a product of nature nonetheless, and not of *ours*.

Now that we have logically concluded that the concept of true-randomness doesn't help the argument of free-will, there is no longer any place for it in the debate between free-will and determinism. We now arrive at the second key term that must be analyzed, free-will.

Free-Will

The term "free-will" is defined by the dictionary in two ways. The first is "the ability to act at one's own discretion." The second definition is "the power of acting without the constraint of fate." Scrutiny of the first definition follows.

In the sentence "the ability to act at one's own discretion," what is "one's own?" If the movement of each subatomic particle determines an outcome, and our minds and bodies are composed of subatomic particles, then "we" are indeed determining our own outcomes. The molecules of our minds are constituents of our minds; there is no reason to assume that our molecules (random or not) must be separated from our identity. As free-will insists that our will belongs to *us*, apart from the laws of physics or nature, it fails to identify what it specifically means by "us." How are *we* separated from the laws of physics or from the program of our functions? Thus, the term "free-will" is incomplete.

At some point, people have somehow established the notion that an automatic process of cause-and-effect strips *us* from *our* state of being, thus believing that only a reality of true randomness may free our will. Perhaps, this false impression is instigated by an incorrect application of the term "random," as people attempt to apply it to the nature of physics rather than to interpret it as our psychological state.

As many sport's psychologists (or psychologists of any genre of mental performance) advise, we perform at our highest ability when we "let it happen," in a state of spontaneity. This is true because we are automatic creatures. Our brains, as the biological computers that they are, run automatically. We think, focus, observe, analyze, problem solve, create, laugh, have fun, cry, grief, etc., etc., etc., all automatically. *We* are doing it, but as the automatic creatures that we are. The moment that we try to fight the automatic process, by *grasping* control over our thoughts and actions, what we are actually doing is *hesitating* or *filtering* our thoughts and actions. This *hold* that we place on our actions, somehow gives us the false impression that we are "controlling" something. When we misbehave as children, our parents tell us to "control" ourselves and behave. However, we are *already* in control, otherwise we wouldn't be functional. When we are commanded to "control" ourselves, what we are really doing in response is *restraining* the actions that we would normally (automatically) express. Over time, we come to the false understanding that the hesitation and tension that we place on ourselves are our means of controlling our situations. Thus, we behave awkwardly during times of high pressure when we believe that the best way to perform is to "control" ourselves; be it while socializing or playing an important match.

As automatic beings, we must understand that the manner in which nature has designed us to control our functions (of *anything*, even thinking or focusing really hard) *is* automatic. It is not supposed to *feel* like we are controlling the process; because to us, when we think we are controlling, we instead perform the act of restraint. It is the term "control" of which many have no real understanding. We associate the feelings of restraint with our intention to control. When we "try" to control, we actually restrain. When we learn that we are too hesitant and tense as a lack of too much *control*, we mistakenly believe that we must stop controlling (opposed to restraining) in order to revert the hesitation. Consequently, when we *try* to not control, we may actually stop functioning and focusing altogether. In this case, the word control is interpreted literally. Our mind's natural sense of control, which happens automatically, is now halted because we believe it is the enemy of our performance, while the true enemy is restraint.

This misunderstanding of the mechanisms of our brains is one of the underlying reasons why so many people struggle to consistently display peak mental performance. The ability of slight restraint, is of course necessary, as our automatic thoughts and actions may sometimes need adjustment. However, many of us are unaware of when and how much we are restraining. Consequently, in some situations we result in restraining ourselves to the point of eliminating all spontaneous thoughts. Because we don't know how to release the restraint, some of us attempt to *force out* our spontaneity, acting out obnoxiously or aggressively.

Many sports psychologists have come to understand that the cure to all of this madness is to psychologically *let it happen* (automatically). Our spontaneity within which our creativity thrives, is the natural state of mind that may be confused for the term "random." *Letting it happen*, opposed to *forcing* (restraining) something, is an automatic psychological state of being. There are many tricks advised to achieve this, such as "being in the moment." Nevertheless, the end goal for mental performance is to free our minds to our brain's natural process, untampered by the restraints that society has conditioned us to use.

As human beings, we feel to be most *ourselves* as we display outcomes that don't seem predictable, rather quite spontaneous. In order to feel like a natural, inexplicable human, we must *feel* random; to have no real explanation for our thoughts or behaviors, because we are unaware of them—we are not overthinking, analyzing and micromanaging them. It is possible that for this very reason, many believe that our free-will must be of a random nature that is unpredictable by any level of mathematics. More reasonably, it may be stated that our 'will' is of a nature that is random to ourselves or to those around us, but not random to physics. Needlessly attacking mathematics, physics and reality will not satisfy anyone's desire to be *in control*. In regards to *free*-will, perhaps the incomplete term was constructed by humanity's interpretation of our psychological state of randomness; later to be confused with an idea that particles must behave randomly even from an all-knowing perspective.

The first mentioned definition of free-will (the ability to act at one's own discretion) presents no greater or different meaning than that of *will* alone. *Free*-will implies that one's *will* is *free* of something, presumably the cause-and-effect function of reality. We have already concluded that randomness would not allow us to achieve a greater control over our will, and so there is no point in insisting on the exclusion of a cause-and-effect reality. "Will" is simply a desire or intent. If our desires and thoughts are automatic, then *we* are automatic, and so the automatic process is a *part* of us. Defining free-will as "the ability to act at one's own discretion" does not free our *will* from a cause-and-effect program. Thus, "free-will," opposed to just "will," still has no distinction. We now approach the second dictionary definition of the term.

Free-will is alternatively defined as "the power of acting without the constraint of fate." And so, we must define fate. Per the dictionary, it means "the development of events beyond one's control." This definition of fate explains why there is an argument between free-will and determinism. Determinism claims that the outcome of events (or fate), including our own actions, are totally in the hands of the laws of physics; and that they are not in *our* control. To further support its argument, it then refers to mathematics and physics to show that all events occur by cause-and-effect. The Free-will side, reactively feeling as though their ability to truly control things is threatened, counters the argument by attempting to destroy determinism's mathematical evidence altogether, claiming that true randomness exists.

The Misunderstanding

Determinism makes the first mistake in this quarrel, by stating that we are not in control of our outcomes. It fails to specifically define what it means by "we." If the actions of our subatomic particles determine the outcome of our events, and *we* are composed of subatomic particles, then *we* are indeed determining our own outcomes.

If, on the other hand, determinism is implying that the definition of "we" that humans formerly held is now eradicated by the concept of predictability, then determinism must ask itself, what does it think humanity's original definition of "we" was? Furthermore, what new definition of "we" is determinism bringing to the table? Could it be implying that the former state of "we" was spiritual, as it is now deemed "robotic" due to the automatic pattern of physics?

Whether our actions are governed by subatomic particles or by a "spiritual process," our actions are nonetheless completely dependent on the program or function by which it exists. We exist in a timeline-format. Whether we have souls or molecules, we will only have one inevitable future; this logic should have been known long before physics discovered the concept of cause-and-effect. By introducing a new idea that things are not random, determinism isn't altering anything about the definition of human control, comparative to the view that humans have always held. Determinism is merely stating that we now know we are predictable creatures.

What has changed? We have learned to predict the weather; do we now say that clouds don't rain because it is actually the molecules within the clouds that cause it to rain? Whether things occur via a molecular process or by a spiritual process, exposing the program which governs the actions does not change the previously known fact that those events are taking place. Exposing that we are automatic does not change the fact that, as always, we control our actions. All determinism really brings to the table is a refining of our original term "random." We may have thought before that things may truly be random, and now it is learned that randomness is only one's inability to predict. This is all that there is to it, though it somehow became a war between science and spirituality. Once more, even *if* processes were random, that wouldn't make our actions any more ours; if something in our minds occurs randomly, how would we take responsibility for it any more than if it were automatic?

Free-will, the poor confused victim in this heated debate, falls prey to the bait imposed by determinism's fallacious argument. Determinism illogically states that we may not control our own actions; rather they are a result of physics. Since *we* are a part of physics, then we (as physical creatures, or objects, or materials, or whatever they want to call us) most certainly do control our own actions. If we *want* (our desire, or will) to put on our shoes, we may most certainly choose to do so. If our desires are a product of physics, then our desires are a product of ours; there is no separation between ourselves and physics. The predictability of nature does not exclude *our* control, it simply implies that we are coherently part of a certain type of program as we function. Determinism is stating that we *do not* have free-will, but determinism doesn't even know what free-will means. There is no definition of free-will; there is no logical explanation or description for it. Therefore, determinism may not say that "something" is *not*, if it doesn't even know what that "something" is. The free-will side, though, errors in this debate by responding to the evidence provided by determinism's illogical argument, instead of scrutinizing the concepts of the argument.

Free-will's illogical reaction to determinism may be an emotional one. Certain subconscious thoughts may haze one's logical interpretation of the arguments exerted by determinism. Without necessarily being aware, free-will activists are in a state of panic. They fear being reduced to the meaningless state of an ordinary object, such as an automatically processing computer or a robot. The fallacy occurring in such thoughts, is that a commonality found between humans and objects is perceived to equate humans to objects entirely. We don't want to be a table, because tables are disposable, not loved, and have no significant purpose in this world. We do not, by any circumstances, want to be equated to such worthless material. As we consider that objects are constrained by the laws of physics, we fallaciously draw the notion that our worth is equated to that of an object's if our thoughts and feelings are also constrained to the predetermined course of nature. To escape this depressing feeling, we argue that we are so much more, since we can think, create art, solve and do so many wonderful things that no "worthless" object, plant or animal may do. We feel a need to protect our identity and worth and to avoid being equated to things that we generally value less than ourselves. As a result, we claim supreme dominance and importance over anything that isn't "us."

At some point in time we then understand that it may be possible to create robots that not only compute and move, but also walk, react to their environments, and even hold conversations! Almost all that can be explained about us is mocked by worthless objects! Free-will activists respond with panic, as the line between objects and *divine* humans narrows. Scientists hint that even consciousness itself may be artificially fabricated.

In order to escape the depressing emotions of derealization and loss of meaning, we search for some magical or supernatural entity within us that no computer, robot, science or anything at all may ever explain or imitate. What results? The inexplicable concepts of "free-will" and "true randomness." If the concepts cannot be explained, then they cannot be mocked or reduced. Thus, if we claim to have these, our identity is safe. The formula works perfectly; the only problem is, the formula is undefined. Free-will cannot be explained, because it is not… a thing.

Conclusion

The war between determinism and free-will doesn't hold one true winner; it holds a confusing battle between two illogically articulated arguments, neither of whom know exactly what they are arguing about.

Determinism brings up a valid point; that reality exists in a cause-and-effect manner. In *any* reality depicted by one solid storyline that is comprised of linked events as one thing occurs after another, there is no escape from the format of cause-and-effect, regardless by what *means* things may occur. If the writer of reality's story is God, and he *randomly* writes events, his "random act" of writing them "happens." The action exists; therefore, it exists and happens in a certain *way*. So even if he himself cannot explain or predict how his storyline will construct (from his subconscious, etc.), it is nevertheless constructed by some definite *means*, rendering the process not random. To say that a storyline is not constructed from cause and effect is to say that a square has no sides, or that a circle has no curves. As long as *events* accumulate into a *storyline*, cause and effect is a concurring property; it is one with the format.

As we humans squirm around looking for different explanations for the manner in which particles move and react in order to escape fate, we waste our efforts as we attempt to fight an idea of something that we don't even understand. There is no reason to be scared of it, it does not imply any negative consequence to our identity or reality. Fate simply means that a reality, which only holds one storyline, is just that, limited to being one storyline. That's all predetermination means. A set future does not take away our ability to control and to act as WE, being one-time-lined cause-and-effect creatures, please. The longing for true randomness is only derived from an emotional need for a fictional reality beyond our understanding, as a response to our insecurity with our identity as limited creatures.

Upon the realization that events are not random, determinism assumes that humans therefore lack the ability to control an outcome, as the molecules that we are composed of ultimately result in our behaviors. The error in this logic lies within the fact that there is no specific definition of the term "human" as used in the statement. If an outcome is a result of the human's body and brain, and the outcome is "not controlled by the human," then what is this idea of a human that is being excluded from the factor of events? If humans are a constituent of the laws of physics, and the laws of physics determine outcomes, then humans influence outcomes. Determinism makes a statement regarding humans, without specifically defining what it means by "humans." In truth, since determinism hasn't identified a part of its vague argument, it doesn't even know what it is arguing about. What is a human, and its will? Can you define any part of a fully functional human being's *will* that plays no role in making any decisions? Unlikely.

Furthermore, the free-will side ACCEPTS this argument as an insult, without fully understanding what the argument means. We know this because, as clarified, the argument doesn't have a complete meaning to begin with. And of course, free-will plays along with determinism's argument to fall down a path of nonsense as it takes on the impossible challenge of proving that things may be truly random, thinking that this is the only means of countering the undefined argument.

All arguments considered, a rational mind may take from this that although every aspect of reality is absolutely definite, humans ought to embrace their identity as automatic cause-and-effect creatures from which a vibrant and beautiful display of spontaneity emerges (spontaneity, at least seen as such by eyes that are not all-knowing).

If, say, in the afterlife (should there be one) humans are all knowing angelic beings, capable of predicting every outcome of the cause-and-effect nature that a timeline is constructed by, one may think that life would become dull. It is difficult to feel suspense or surprise as one watches a movie for the second time. Well, does one not enjoy their favorite dish, despite knowing how it will taste? Such all-knowing beings may be focused on the pleasant stimulations occurring in the present moment, wasting no time with thoughts of anticipation or dread, in just the manner as we ought to be living our lives here on Earth! No beauty or glory is eradicated by the mere introduction of event-based mechanisms.

On the other hand, one may claim that it is unfair to be imprisoned by a certain path or fate without control over the whole storyline that is "written" for us. Although some may feel disappointment in discovering that one may not completely write their own story line, humans have *always* known that we cannot choose many aspects of our own lives. For starters, we cannot control the manner in which we are born. Our existence is entirely at the mercy of the universe (or God) from which it began. We may be born into any situation, killed, or harmed in an infinite number of ways that are out of our control. This is nothing new to us. Fate, is just a description or a recording of what events take place. Whether events are determined or identified *before* or *after* they take place, is irrelevant to anything other than our ability to predict what will happen. It bears no weight on the discussion of our will, or control.

That said, it is this one grand storyline that gives any being the chance to exist in the precise manner that it does. The fact that we are given any opportunity at all to feel any sort of stimulation or experience, should provide us with a sense of gratitude and thrill. Yet, as humans, we always seek more. We were okay with lacking all control whatsoever over choosing our beginning (birth), since that has already passed; but now, we want the ability to dictate any desired outcome of our storyline without any resistance applied from the universe, as if we have the right. Unfortunately, unless we are some sort of god, none of us have the power to limitlessly control what we are or will be, any more than we may control by what means we are born.

It is with gratitude and humility that we must accept the uncontrollable and control what we may. Our fate, or outcome, is inevitable. What we *can* control are our actions, in just the same manner that we have always controlled them. Defining our process of control with a predictable molecularly functioning one, does not strip us of control.

Although we exist in a predetermined, automatically sequenced universe, *we*, as the automatic creatures we are, each play a role in developing life's grand storyline. Our cause-and-effect nature doesn't stop us from laughing, loving and losing ourselves in the beautiful stimulations and wonders of our reality.

Chapter 9:

Philosophy & Reality

Western literature, up until the late 1800s, commonly expressed views of realism; a depiction of reality oriented in such a manner as to reveal things for what they truly and objectively are. The modernist movement in literature struck the early 1900s with philosophical perspectives questioning our ability to perceive the objective reality. Furthermore, it even questioned the existence of an objective reality altogether. It was at this time in 1905, when Einstein developed his theory, claiming that motion itself has no objective (absolute) value. Although the mentality of modernism remained in fashion for a good thirty years, it was not by any means the first of its appearance. Philosophy has long questioned certainty. *Hippias Major*, a dialog written by Plato around 350 BC, tells of an expedition set out by Socrates and Hippias in search of a definition for the term "beauty." No one would expect such a familiar and common word that has long been intuitively understood, to be undefined. Socrates and Hippias result in failure, as they discover no one formula (definition) which coherently applies to all uses of the term. If Anna is beautiful from Bob's perspective, but not from Kyle's point of view, then how do we determine whether or not Anna is beautiful, by definition?

Philosophy is the Question, not the Answer

Philosophy is the beauty of question and doubt—it questions society's certainty of a topic that may *seem* clear, as it may be very familiar, to prove that we know very little. Any element in our lives with which we are constantly associated becomes our *norm*, and so we feel as though we truly know and understand it. This is a result of familiarity, not necessarily comprehension. A philosophical question points at such familiarities to test our true understanding.

Far more importantly than a mere hobby or interest, philosophy is an intellectual mechanism that should be present within all of our thoughts. Just as we don't want our home computers to lack a hard drive or a processor, we should not want for our brains to be lacking this crucial manner of thought. Just how important is it? It isn't just a subject you sign up for in school that you may never use. Philosophy marks the difference between a reasonable and a horrible, gruesome culture. Humanity's ability to question the norms is anything but unnecessary.

In the 1700s, did the average child question the norms as their parents, all households on the block, along with the rest of their country owned and abused slaves? From the moment they could walk, or talk, such settings have existed in their backgrounds, no less common than their tendency to breathe air or drink water. They were not genetically *bad* people by nature; it is just how the human brain (biological computer) functions. Norms are accepted or ignored, new information is accepted as received at first impression, while the altering of current information (change) is resisted unless there is a powerful reason for the change. With the start of slavery, its introduction to the human brain went somewhat as follows...

Seller: "Hey guys, here are these slaves!"

Buyer: "What are slaves?"

Seller: "These workers that are *meant* for working for you."

Buyer: "Are we allowed to do that to people?"

Seller: "These *people,* are not like you and me. Look. They look different; they were *meant* for this. They are property."

Buyer: "Oh, okay."

The new information (fallaciously presented, of course), especially when it benefits one's quality of life as many others are supporting and encouraging it, becomes the norm. *Change* to a norm is naturally resisted, *especially* when the norm is a large factor of one's quality of life. It is why so many African Americans were required to fight so hard and sacrifice so much for the change to occur. It is only a result of the manner in which the biological brain functions as it attempts to survive and thrive. A lion will not discontinue consumption of its prey unless a better and more convenient source is offered.

The White Americans of the 1700s simply functioned like the natural animals they were. Unfortunately, humans (of all ethnicities) are very much the same today. We do not continue to enslave, because the change has occurred with a gradual transition into a new norm along with a new set of standards. Many of us continue, however, to be the same slow-changing, absent-minded, sheep-like followers that we've always been, so easily shaped and influenced by societies constructs and opinions.

Change for the better should not require so much time, effort, sacrifice and *violence*, considering that we have the tools necessary to bypass it all. One of these tools is philosophy. It questions our norms, as familiar and correct as such norms may feel. This is what evolves the animal. It is what clears the fog, and welcomes reason and clarity.

Philosophy is the question; to pave way for a journey to the right answers. A philosophical mindset, however, is not the *end-goal*. We tread in search of truth with an open-minded method of learning to discover just that, the answer. At some point, one must return to reality with a conclusive response for the abstractedly constructed question. The process required to derive a reasonable answer is critical observation, examination and a logical method of defining the key concepts involved.

Narrowing Down the Answer

Now, we narrow down the answer to Socrates' and Hippias' question... We will scrutinize the term to the core to develop a universally applicable and coherent definition.

What do we *feel* when we call something beautiful? We feel pleasure. Thus, beauty is a *feeling*. When, specifically, does the pleasure occur? When we perceive some sort of information.

Thus, beauty is simply, "pleasantly perceived information." *Information* includes (though it has no limits) visual, audible, physical, conceptual, tasteful, and any other form of information that may be perceived by one in any manner. "Beautiful" is the characteristic of any one or thing from which information is pleasantly perceived. Although one person may not

112

be perceived as "beautiful" by every observer, this definition of beauty allows the term to be held relatively to the observer's perception, rather than to be held solely by the noun that is being perceived.

If Jessica loves Kyle (who is perceived as ugly by most who encounter him), then Jessica may have visually associated Kyle's unattractive face with great feelings of pleasure. Every time Jessica looks at Kyle's face, she is nothing but pleased as she melts in her shoes. Thus, to Jessica, Kyle's face is extremely beautiful. This is not a matter of being "nice" to Kyle. It is not a cliché, romantic pick-up line. This is pure logic and truth. May Jessica logically analyze Kyle's facial structure to determine whether or not it possesses pleasantly interpreted characteristics (such as symmetry)? Of course, she can; and she may logically verify that Kyle's face may not have beautiful structural or textural qualities. Jessica may even have mixed feelings about Kyle's face! She may at times be pleased, and at other times be displeased by his facial features, depending on her mood or the perspective from which she is interpreting the information (logical perspective, emotional, etc.). This would mean that Kyle's facial information is received variably; at times as *beauty*, and otherwise at other times. Jessica may even find Kyle's specific structural features to be pleasant by association to her feelings of love, despite most of society interpreting Kyle's features unpleasantly. If information is pleasantly interpreted, it is thus beautiful in that instance. If the information is interpreted in another way, or by another person in a manner that is not pleasurable, then it is not considered beautiful in those specific circumstances. And so, we have logically constructed a coherent definition for the so-called ambivalent concept of beauty.

Socrates' and Hippias' reluctance to accept the original definition of the term was a wonderfully logical approach, because the original definition was not applicable to all circumstances. However, an answer will always exist, so long as the question is specific, and the topic is sufficiently observable. If a term finds no definition, then the dynamics of that term must be more specifically understood. In this case, realizing that beauty is a *feeling* led us to the answer. What do we *feel* when we call something beautiful? We feel pleasure. When, specifically, does the pleasure occur? When we perceive some sort of information... We identify the source of the meaning (feeling), and then we identify its dynamics; that is, the manner in which the term applies or does not apply to varying situations.

Einstein's Brilliant Philosophical Question

Much like Socrates questioned our interpretation of beauty, Einstein asks the question, "what is motion?" As mentioned, an answer may be found, so long as the question is specific. Einstein not only asked the question specifically; he asked it brilliantly. His question was even followed by obstacles as to ensure that it isn't easily dismissed or brushed away with weak replies. He explains how motion may only exist if there is a reference point to *call* it motion. Although motion has always been intuitively accepted as *certain*, he validly points out that we are blind to certain aspects of this topic. We may be moving through space along with our galaxy at a tremendous constant speed without even feeling it. Our blindspot is that we cannot detect motion that is of unaltering speed or direction. Einstein's question, stops here.

Before anyone invests a great deal of effort towards answering this question, Einstein proceeds with the assumption that motion simply does not exist. Postulate 1 takes advantage of our blind-spot in constant motion, stating that stillness must be considered equal to constant motion (since we cannot tell the difference). This *rule* is asserted only because we could not disprove it. Applying a principle of light together with Postulate 1, Einstein develops a theory that displays sporadic results, allowing two contradicting events to occur simultaneously. He uses our lack of knowledge of motion to break and crumble reality as we know it. Meanwhile, physicists lose sight of the original question, assuming that it is simply unanswerable, much like Socrates and Hippias agreed that there is no definition of beauty.

His Question Even Begs Us to Answer it!

His theory serves as a riddle to call forward an answer that may only be accurate if a reference point for absolute motion is discovered. Ironically, if one looks with enough depth, the riddle is formatted in a manner that points towards the direction in which to search for the answer. The motionless reference point hides within his very own theory. The element which causes chaos in scenarios constructed with Special Relativity's rules, is the very same element with which the theory is answered. In *light* there is truth, which brings clarity to our understanding of motion, space and time. If Einstein could not *see* true motion, then perhaps he should've looked in the light. Well, maybe that wasn't Einstein's job. He's the one who applied forth the work and effort to lay out the important question. WE were to try and answer it.

Relativity assumes that the direction of light may be influenced by the motion of its source perpendicularly to its direction of emission. No explanation is offered to support the behavior of this fallacy. It is almost as if Einstein knew the answer to his question, and left us a hint to discover it. *"Look guys, this light over here, which is moving sideways (which it shouldn't) is causing CHAOS to occur. A-hem! Anyone?"*

Yet... We Look Elsewhere

It is possible that all who have allowed this fallacy to slip by, did so with the preconceived notion that *any* thing that is thrown, shot or propelled vertically, will simultaneously travel horizontally if the motion of its source is horizontal. The simple fact was somehow overlooked that photons do not have mass, and that waves do not behave in such a manner. Inevitably, Mazi's scrutiny of the theory of Special Relativity has directly evoked the development of the opposing theory, the Photonic Reference Point.

Albert Einstein was a brilliant mathematician. The concept on which his theory and mathematics are based, however, is purely philosophical. "Nope, you're not moving. Prove it!" Indeed, it isn't easy to prove. One hundred years pass, and no one can say "I'm moving!" Philosophical questions have an interesting way of making people feel stupid. It questions people's certainty by inquiring specific answers that they aren't accustomed to finding.

Most people think and process intuitively, without pondering with a great deal of depth about why things are the way they are. We say "we just know." Philosophy is annoying. It isn't someone to go out on a typical date with. It asks us what we know. It then asks us *how* we know. It even asks us *why* we think we know what we know. It might even ask us if we know who is asking the question (just kidding). As annoying as it may seem, it is good for us. We are accustomed to accepting things without giving them much thought, *especially* if everyone else is on board. It is with practice, though, that one may become quite good at answering such brain-teasing questions, or even constructing new questions.

Einstein's question was a philosophical beauty. Finding an answer to it, is also a beautiful process of scientific development. The science community faulted, however, by accepting his theory the moment it was known that there was no clear answer to the uncertainty. It is doubtful that science has spent a great deal of effort in discovering a true reference point to answer Einstein's question. If anything, most of their investments went squandering within the testing of his theories and formulas, which were of course aligned; as he wouldn't have constructed them in a manner otherwise. Had deeper thought been applied towards the philosophical region of his work, who knows how early his fallacies would've been discovered, and his question answered.

Einstein's Angle

Let us assume for a moment, that Einstein did not actually believe in his theory when he plotted it in the early 1900s... Around that same time, the hotel-riddle (referred to as The Missing Dollar Riddle) was written and based on other math riddles developed in the 1800s. Let's say that Einstein had come across this riddle. He observes how others around him respond to this riddle, and to his amazement, they were not able to solve what should've been a cake-walk. Even for him, the seemingly easy math problem may have required a bit more effort than he had expected. Perhaps it required him 2 seconds to inhumanly solve, rather than the 0.01 seconds of thinking that he would normally expect himself to solve such a problem involving only dollar bills. Nonetheless, he is intrigued by the ease with which people are deceived by such an elementary math problem. He notices that people seem so preoccupied by the mathematics involved, that they completely miss the fallacious inferences. Recalling back to Chapter 1; the hotel-riddle confuses the audience by using a number in a manner that *seems* appropriate, only because it vaguely belongs to a particular group. This fallacy is constructed by an incredibly simple method of deception; simply assigning a number to a similar group to which it does not belong. Yet, it proves to be an extremely powerful method of deception when weighed against the general person. A simple fallacy, in a *very* simple math problem (30 − 3 = 27) is able to leave fully-grown, educated scholars speechless.

Einstein wonders, then… "What if I were to create a similar riddle, containing complex theoretical physics that runs down for pages? How preoccupied would the reader be with the mathematics then? And for the fallacy? Assigning a number to an incorrect meaning? No, too easy! Someone will eventually discover the fallacy. Instead of applying a very simple fallacy, why not employ a great philosophical question that cannot be answered? Humans already struggle with philosophy as it is; to employ one that is impossible to answer, within a mindboggling math and science problem… This could work!"

And so, he asks, "What is motion?" The philosophical question alone already has all of us scratching our heads. While we're still trying to wrap our heads around how we know that constant motion is different from stillness, he has already moved ahead of us, claiming that motion and stillness are exactly the same thing, since we cannot tell a difference, physically. He walks us through scenarios whereby two observers, which are *moving* relatively to one another, are *both* assuming that they are *not* moving! So, two people, who's distance between one another is changing, are *both* claiming that they are not moving! Einstein then says that they are *both* correct, since constant motion is the same as stillness. Then, light comes into play. Einstein now adds more rules to completely scramble our minds. Light's speed of travel is witnessed equally by all observers, since all observers may claim that they are motionless. From Eve's perspective, since she may claim that she is motionless, Jack is moving. Since Jack is moving, his laser must be traveling sideways along with him, meaning that it travels a greater distance. Given that it travels a greater distance, considering that its speed *must* be the same for all observers, it is now requiring more time to complete its path of travel, relatively to Jack's perspective. Already, most people by now are completely lost! What in the world is going on?!

There are so many holes and unanswered questions as the theory moves on at the speed of light. At this point, Einstein's "riddle" is purely philosophical, and it is scrambled with all sorts of fallacious assumptions. The aspect of the theory that serves as a philosophical riddle, alone, deserves a Nobel Peace Prize as it is cleverly constructed in a manner that deters physicists away from the original question. They forget that they *must* find a motionless reference point in order to refute its logic. Since no such reference points are apparently obvious at first glance, they leave it alone.

To make matters worse, the brilliant mathematician that Einstein was, devises formulas to adapt his theory of insanity into physics. Just as talent varies amongst computer coders and programmers, as only few hackers possess the talent to break into heavily secured networks, Einstein was an incredibly talented formula writer who could adapt a theory to a similar version of our physics laws. Note, he didn't adapt his theory into the formerly understood physics, he actually altered some of physics in order to fit his theory into place. Now, his riddle "makes sense." It is supported by mathematics. At this point, physicists only double-check the mathematics which are bound to Einstein's supposed rules of reality, closing the doors to his philosophical question for good.

Einstein has taken the hotel-riddle, and used electromagnetic waves to radiate it into an incredible hulk. Fully grown adults were too preoccupied with subtracting 3 from 30 and adding 2 to 27 to notice that "2" was improperly categorized. Einstein's hulk-ified hotel-riddle now leaves physicists too preoccupied with pages of theoretical physics to notice "something" about a philosophical question that no one can wrap their minds around. Unreal.

Einstein gains acknowledgement as one of history's greatest geniuses, to enjoy the remainder of his life in fame as the one who revolutionized physics. Perhaps, when his theory is finally proven wrong, a letter from him to the world is revealed. And what does it say? "Gotchya!"

Well, what a sick joke that was. How much money was spent testing his theory and studying his work? Perhaps he only ever wanted someone to find the motionless reference point. Or, maybe he intended to *hide* the motionless reference point, knowing that such knowledge could potentially lead to war weapons of mass-destruction. Perhaps, Einstein had solved time-travel by utilizing knowledge of the motionless reference point; knowing such power could fall into the wrong hands, he may have developed his theory to obscure science from ever discovering the secret. Well, if this is true, then Einstein is an epic hero; and as the author of this book, I am quite the numbskull for throwing away years of his heroic work and placing all of humanity in danger, only to correct a misconception that bothers my obsessive compulsion disorder.

Physicists' Overexcitement with Philosophy

Jokes aside, it appears these days as though the world of physics is philosophically trigger-happy; overly excited to strongly consider the craziest out-of-the-box ideas possible. Rather than considering the possibility that a photon and an electromagnetic wave are two separate entities (as a photon surfs within an invisible wave), science today is instead enthusiastic in stating that a single photon behaves as both a particle and a wave, traveling through both slits of the double-slit experiment simultaneously and interfering *with itself,* but *only* while it is not being observed.

As if the interference pattern resulting from the experiment doesn't count as undergoing observation; physicists believe that the direct observation of a photon only counts when a detector is specifically watching the slits to see through which hole the photon passes. Logically, there is no consistency with this structure of thought. A photon's resulting interference pattern is just as much an observed quality of the photon as is the direct detection of the photon traveling through the slits. If one is to assume that the photon will *know* it is being observed when a device is detecting its motion through the slits directly, then why would one not expect the photon to feel self-conscious when the back screen displays the photon's behavior? Rather than concluding the more logical assumption that the device set at the slits is somehow interfering with the behavior of the photons, physicists jump to wild philosophical ideas involving the observer's weight on the course of reality.

Neither does Einstein's theory do logic total justice. Rather than stating that his equations *assume* motion to be relative for the sole purpose of his equations to apply, Special Relativity assumes that objects simply cannot pass through definite points in space, which is what absolute motion would essentially be. To infer that absolute motion does not exist only because a limited observer cannot *detect* true motion is short of logic. Why must one *sense* true motion in order for it to exist? Why may we not *deduce* its existence? If separate points in space exist, then the idea of one object passing from one point to another may be considered to deduce the existence of true motion. Einstein's philosophical question may have led to some interesting discoveries of time variations caused by motion or gravitation, but as mentioned earlier, at some point we must return to reality with a definite explanation. Recalling back to Chapter 5 of this book, these effects may be explained in a realistic manner, without abandoning our long-standing logical beliefs of an objective reality. An object's particles may slow down as its speed through space is increased. An object's particles may also slow down as a gravitational force is exerted onto each of the object's constituents. Both of these effects were discovered with thanks to Einstein's philosophical question.

As science rapidly increases in complexity, new questions arise that are not easy to answer. Einstein asks, "how can we prove the existence of true motion?" The double-slit experiment asks, "why do single-fired photons produce interference patterns in such manners?"

What ever happened to... allowing a mystery to remain a mystery until it is solved? Physicists are so quick to answer unanswerable questions with any random outrageous answer that comes to mind. To Einstein's riddle, they answer "true motion actually doesn't exist, its subjective!" To the paradox of the double-slit experiment, they conclude that "the photon is playing with itself, but only when we're not looking! Its existence and location are subjective!" Until something slaps us in our face with undeniable proof, we should not ever abandon logic as new obstacles present themselves. The rules that humanity have held thus far, such as "one thing cannot be in two places at the same time," have long remained rules for a good reason—our experience has taught us such values. To abandon common knowledge out of excitement for something new and sporadic, is quite frankly immature.

It may have been Einstein who gave physicists the courage to defy the normal views of reality that we have always believed in. Ever since he had devised a good (rather tricky) argument to eradicate an idea of motion that we have so long intuitively believed, scientists have grown the cojones to make all sorts of absurd statements. You may have heard the new question that physicists are tickling themselves with... "Does anything really... exist?!"

There is a structure of logical analysis that theoretical physicists ought to abide during examination of any topic. You may have noticed while reading through the chapters, that the underlying train of thought was a method of scrutiny. In comprehending the concepts at hand, we inquire of their capabilities, actions, tendencies, mannerisms, and limitations. It is in this manner that matters of any topic may truly be understood for what they are. If they cannot be understood, then they shall remain unsolved. One should not, however, jump to radical conclusions to abandon the current understanding of reality; at least not without undeniable evidence.

It is my hope that humanity (myself included) will improve its ability and tendency to objectively analyze and comprehend concepts with greater discipline. Perhaps, in doing so, it may reduce the imaginary conflicts that it creates for itself (such as the dispute between free-will and determinism, whereby neither party understands what they are even arguing about). With more clarity and understanding, comes peace and progression.

The Author

I do not have, but a gram of education beyond high school; I am quite limited in knowledge pertaining to science. I cannot recall more than twenty percent of the content from the table of elements. Upon first hearing of Einstein's theory, I was shaken by the extent to which it seemed counterintuitive. Visualizing space as a grid, I could not understand why a scientist would not be able to visualize an object moving through different points along a space grid (which is what absolute motion essentially is). Determined to prove the existence of absolute motion, I began to scrutinize Special Relativity to understand *why* it abandons the concept of true motion. As my first question led to my next, my curiosities would guide me to the findings expressed in this book. My tendency to investigate logical aspects of the key concepts at hand is, by far, my favorite intellectual endeavor. For this habit, I must give major credit to my elder brother. Growing up, his personality (as annoying as it may have seemed) held me accountable for nearly every thought and statement I would ever make. "Why, but why, but I thought you said that, but why this, but why that?" His unique personality, which constantly demands for scrutiny, had forced me to find an answer and explanation for even the most subjective statements. From a very young age, we would often debate for hours upon hours, scrutinizing the smallest of arguments, and dissecting them into hundreds of little pieces. Over time, it has become habit to *attempt* to understand the truth of matters with a solid method of scrutiny and reasoning. Although our debates were often heated and aggressive, it was the competitive essence which fueled my desire to keep at it. For such experiences of intellectual stimulation and development, I will forever be indebted to the relationship I've held with my elder brother; and I encourage others to challenge one another in like-manners.

Author's Theories

1st THEORY: PHOTONIC REFERENCE POINT

Given that a photon is massless, it may not accrue momentum from the physical motion of its source. Therefore, a photon's velocity in the perpendicular direction relatively to its direction of emission must be absolutely zero. This motionless dimension of a photon serves as a reference point which indicates the stationary framework of space and at the very least, the absolute motion of any object or thing. Leaving the direction of emission aside, the direction of travel, too, indicates a photon's motionless state perpendicularly to said direction.

2nd THEORY: SPACEPOINT COHERENCE

No object may simultaneously exist in multiple places at any instant, nor may multiple objects exist within the same exact place at any instant. Respectively, no single point in space (an available space within which objects may exist) may contain multiple objects at any one time, nor may multiple points in space contain the same one part of an object at any one time. Therefore, all points in space (or spacepoints), of which there are an infinite abundance, must each be regarded as individually unique entities and coordinates of space. That said, the distinction between the constant motion of an object and the motionless state of an object is indeed apparent to the laws of physics, as it determines which spacepoints are occupied by an object, and which spacepoints are not occupied by an object at any given time.

Author's Theories

3rd THEORY: ABSOLUTE TIME

Each spacepoint is bordered by other spacepoints, as they allow for information (particles) to be transferred between one another based on the set of formulas (characteristics) by which the information may behave as it is transferred or stored. The interaction between spacepoints are synchronized in such a manner that they do not allow for contradictions to occur, such as multiple pieces of information being present in one location or one piece of information being present in multiple locations within the same universal instant. (Universal Instant: The relative position of all objects and energies in existence at one given moment). This limited manner by which space is synchronized, is time. Time is independent of the presence of objects, and only dependent on the formulative synchronization of space whereby particles or energies "may" be stored or transferred in a non-contradictive manner between any two given universal instants. A change may not occur without time; therefore, time is the allowance of change of information to occur in a manner limited by the synchronization of space. Time allows an infinite number of increments of chronological change in space. As time continues to allow such change in space, the universal clock continues to run, whether change is occurring or not. Conclusively, time is space's synchronized sequential allowance of non-contradictive change to occur.

Author's Theories

4th THEORY: WAVE-PARTICLE DISTINCTION

Explanation to Double-Slit Experiment...

Given that each photon emission causes no more than one dot on the screen, we must state the fact that there is only one photon particle involved with each emission. Separately emitted photons exclusively meet the screen at which points a wave interference pattern in this scenario would meet the screen. Given this fact, we must conclude that two waves are indeed exiting the slits simultaneously, since the displayed interference pattern requires the interference of two waves which exit the two slits. The passing of the two waves through the open slits suggests that one larger wave met the plate before separating through the slits.

Thus far, we know that each emission involves one emitted photon and one emitted wave towards the plate (whether or not they are the same thing). We also know that one photon and two waves exit from the open slits from the back of the plate. We know that the photon may only pass through one slit, since one thing may not exist in two places simultaneously. Thus, from each single photon emission, we know that each slit allows one wave to exit, and that only one of the two slits allows one photon to exit. These facts suggest that the photon and the wave emitted are two separate entities.

Since the two waves exiting the slits (after the single wave is divided by the plate) do not imprint a wave pattern on the screen, we may conclude that the wave(s) are invisible and undetectable by the screen. Considering that only the photon is visible (imprints the screen), and that each emission allows

Author's Theories

... 4th THEORY: WAVE-PARTICLE DISTINCTION

... a photon to meet the wall only at which points the interference pattern suggests that photons may be struck, we may logically assume that a photon is traveling in a manner suggested by the invisible waves.

To comprehend this proposition with ease, one may visualize that on some given point within the invisible wave (before the plate divides the wave into two), a single photon particle is permanently stationed, as if "surfing" on that certain point of the wave. As said point of the wave interferes with other waves, the wave alters, accordingly altering the location of the photon. Where the wave ceases to exist due to wave interference, the photon cannot surf due to a lack of wave. Consequently, each emitted photon may only meet the screen at which points are met by the invisible wave. As many emissions occur at separate times, each of the photons surf on certain points of their own waves; the position of each photon relatively to its wave will vary from photon to photon, as the specific subatomic location from which point the photon is emitted varies. Each photon may only meet the screen at which points the waves tend to make contact. Each emitted invisible wave is identical to all previous and following waves, as waves are emitted from their source spherically in all directions, and both the location of the source and the slits remain unaltered; therefore, each wave will meet the screen with the same identical interference pattern. Conclusively, electromagnetic waves and photons are two distinct entities. When a single photon is emitted, an invisible electromagnetic wave is distributed in all directions, within which the photon surfs on a certain point of that wave. Since electromagnetic waves transport photons, and electrons absorb photons, then it follows that electromagnetic waves may also move electrons, hence the results of fired electrons in the double-slit experiment.